S0-AKY-180

Reactivity and Structure Concepts in Organic Chemistry

Volume 31

Cornelis Blomberg

The Barbier Reaction and Related One-Step Processes

With 4 Figures and 49 Tables

Springer-Verlag
Berlin Heidelberg New York
London Paris Tokyo
Hong Kong Barcelona
Budapest

Professor Dr. Cornelis Blomberg
Henk Oostveenstraat 10
1391 EX Abcoude
The Netherlands

ISBN 3-540-57169-8 Springer-Verlag Berlin Heidelberg New York
ISBN 0-387 57169-8 Springer-Verlag New York Berlin Heidelberg

Printed in Germany

Typesetting: Macmillan India Ltd., Bangalore-25
51/3020/SPS – 5 4 3 2 1 0 – Printed on acid-free paper

List of Editors

Professor Dr. Klaus Hafner
Institut für Organische Chemie der TH Darmstadt
Petersentraße 15, 64287 Darmstadt, Germany

Professor Dr. Jean-Marie Lehn
Institut de Chimie, Université de Strasbourg
I, rue Blaise Pascal, B.P. 296/R8, F-67008 Strasbourg-Cedex

Professor Dr. Charles W. Rees, F.R.S. Hofmann
Professor of Organic Chemistry, Department of Chemistry
Imperial College of Science and Technology
South Kensington, London SW72AY, England

Professor Dr. Paul v. Ragué Schleyer
Lehrstuhl für Organische Chemie der Universität Erlangen-Nürnberg
Henkerstraße 42, 91054 Erlangen, Germany

Professor Barry M. Trost
Department of Chemistry, Stanford University
Stanford, CA 94305-5080, U.S.A.

Professor Dr. Rudolf Zahradník
Akademie der Wissenschaften der Tschechischen Republik
J.-Heyrovský-Institut für Physikalische Chemie
Dolejškova 3, 182 23 Praha, 8, Tschechische Republik

To my wife

Preface

The significant increase in applications of Barbier and Barbier-type reactions in (in-) organic synthesis over the past twenty-five years, combined with repeated stimulation on the part of Springer-Verlag (Dr. R. Stumpe) were strong incentives for me to write this book.

My earliest activities in this field include a modest and first-ever review article (in cooperation with Dr. F. A. Hartog who at that time (1977) did his doctoral research on the Barbier reaction). References to this review article, as mentioned in the Science Citation Index till recently are a good indication for a continued interest in this simple synthetic procedure.

The subtitle of this monograph, *Related One-Step Processes*, may require the following clarification. A synthesis with the aid of an organometallic compound is usually carried out in two steps: 1) the organometallic reagent is prepared from an organo halide and a metal in a solvent, 2) a chosen substrate is added to this reagent. In a Barbier reaction these two steps are combined: a mixture of the organo halide and the substrate is added to a suspension of the metal in an appropriate solvent.

It is clear at first sight that this simultaneous addition of two reactants to a metal will lead to a sequence of steps in the reaction mechanism. The *one-step* character of the Barbier procedure, as mentioned in the subtitle of this monograph, is therefore *not* to be found in its mechanistic aspects but in its simple technical performance. It is essentially this elegant procedure that has made the Barbier reaction so attractive for many applications.

I hope and expect that this monograph will be helpful in drawing the attention of synthetic chemists even more strongly than it was in the past to this slightly 'overlooked' reaction that has become more and more attractive, not least because of the selective use of several different metals and of a growing understanding of its multi-step mechanism.

I wish to express my gratitude to Dr. Greg Hitchings for his meticulous readings of and his proposed corrections in one of the early versions of the manuscript.

Amsterdam, The Netherlands
September 1993

Cornelis Blomberg

Table of Contents

1. Introduction

1.1 Magnesium: A New Metal in Organic Synthesis

One of the great and important steps in the history of modern chemistry was made by Philippe Barbier [1] when, in 1899, he started his work in the area of organomagnesium chemistry. In doing so he discovered the one-step reaction that was to be named after him; the Barbier Reaction.

This reaction was to break away from the zinc-tradition that had dominated the use of metals in preparative organic chemistry until the beginning of the twentieth century.

Part of this zinc-tradition was the reaction of an organic halide with metallic zinc in the presence of a given substrate. After the appropriate work-up of the reaction mixture the desired product was then isolated:

$$\text{organic halide} + \text{Zn} + \text{substrate} \xrightarrow[\text{2) work-up}]{} \text{product}$$

Barbier saw new possibilities for this type of reaction with magnesium instead of zinc. His introduction of this metal to preparative organic chemistry opened up a new area in the history of chemistry.

1.2 The Early History of Organozinc Chemistry

1.2.1 The Beginning of Organozinc Chemistry

Frankland; 1849. The zinc-tradition in preparative organic chemistry had its start when Edward Frankland [2] discovered the formation of alkylzinc compounds from the reaction of iodoalkanes with metallic zinc:

$$\text{iodoalkane} + \text{Zn} \xrightarrow[\text{sealed tube}]{\text{no solvent; 150 °C}} \text{alkylzinc compound}$$

Frankland carried out this reaction in the expectation that organic radicals could be generated by the removal of the electronegative halogen atom from the haloalkane by the more electropositive zinc atom.

In his first publication of this work he reported that when iodoethane and metallic zinc were heated in a sealed glass tube for two hours, a colourless mobile liquid and some white crystalline material resulted.

Large amounts of gas also escaped on opening the sealed tube and Frankland reported his results of its analysis.

Probably in an attempt to trap 'radicals' which otherwise would escape through sidereactions, Frankland also performed this experiment in the presence of some reagents such as water, ethanol or diethyl ether.

It is worth noting that he reported that little gas escaped on opening the sealed tube from the reaction mixture to which diethyl ether had been added.

In a second publication on the same subject in 1849 [3], Frankland reported that when iodomethane was employed as the halide, a white crystalline material resulted from which a colourless, transparent liquid could be distilled when heated in a hydrogen atmosphere. The high reactivity of this liquid, which burned spontaneously when brought into contact with air, led to Frankland describing it as

very likely, that the peculiar compound, which I propose to call methylzinc, is capable of playing the role of a radical and to react with oxygen, chlorine, etc . . .

Frankland; 1855. Diethylzinc Purified. From his third publication in 1855 [4], it appears that Frankland had several unpleasant experiences with the preparation of the alkylzinc compounds. He made mention of a new apparatus, constructed by an engineer, in which he could easily prepare 4 to 5 ounces of diethylzinc which he mainly used for analytical purposes. Furthermore it is worth noting that

a. diethyl ether was used again as a solvent (same volume as iodoethane),
b. the reaction temperature was 150 °C, and that
c. pure diethylzinc was obtained by distillation at 118 °C.

The results of Frankland's analyses of this compound were as follows:

	Calculated	Found
C_4H_5	47.14%	47.32%
Zinc	52.86%	52.67%

In this publication Frankland also reported the use of diethyl ether as a diluent in reactions of ethylzinc with iodine, bromine, sulphur and water; without it these reactions were reported to proceed too violently.

1.2.2 Improvement of the Preparation of Diethylzinc

Frankland; 1849 and 1855. Thus Frankland had noticed the advantage in using diethyl ether as a solvent for the reactions involving metallic zinc. Furthermore, in 1849, he reported that sidereactions were suppressed by the use of this solvent [2]:

Equal volumes of iodide of ethyl and ether were heated in a sealed tube with zinc, to a temperature of about 150 °C (302 °F), *until the action appeared complete: on being allowed to cool, the residual thick, oily, fluid did not solidify. When the beak of the tube was afterwards broken off, only a few cubic inches of gas were evolved*, . .

Also in a later publication, in 1855, Frankland noticed the advantage of the use of this solvent [4].

Pébal; 1861. A further six years passed before Pébal [5] discovered the general method for the preparation of organozinc compounds. Instead of sealed tubes and by using diethyl ether as the solvent alkylzinc reagents could be synthesized in ordinary laboratory glassware. He demonstrated this for the reaction of iodoethane with zinc which had first been activated by reaction with sulfuric acid.

Rieth and Beilstein; 1862. One year later, Rieth and Beilstein [6] reported a facile and productive synthesis of diethylzinc by reaction of iodoethane with a 4:1 zinc/sodium alloy without solvent under an atmosphere of carbon dioxide.

After a few hours the diethylzinc could be isolated quantitatively by distillation directly from the reaction mixture.

1.2.3 Synthetic Applications of Organozinc Compounds

Freund; 1861. It was Pébal's co-worker Freund [7] who was the first to apply the newly developed organozinc compounds for synthetic purposes in organic chemistry. He prepared several ketones by the reaction of these organometallics with acid chlorides.

Freund reported the following equation to show his reasoning for the replacement of chlorine by an alkyl-group:

$$\left.\begin{matrix} C_2H_3O \\ Cl \end{matrix}\right\} + \left.\begin{matrix} CH_3 \\ Zn \end{matrix}\right\} = \left.\begin{matrix} C_2H_3O \\ CH_3 \end{matrix}\right\} + \left.\begin{matrix} Zn \\ Cl \end{matrix}\right\}$$

Chloracetyl Zincmethyl Acetyl-methyl Chlorzinc

The organozinc compound was first prepared in a sealed tube with diethyl ether as the diluent; after this reaction was finished volatile material was distilled into a flask after which acetyl chloride was subsequently added dropwise.

Surprisingly Freund did not make use of Pébal's procedure for the preparation of diethylzinc in ordinary glassware. Using dimethyl- and diethylzinc

Freund was able to synthesize what was named 'acetyl-methyl' (propanone), 'acetyl-ethyl' (butanone), 'propionyl-ethyl' (3-pentanone) and 'benzyl-ethyl' (1-phenylpropanone).

Rieth and Beilstein; 1863. From Rieth and Beilstein's publication [8] it is difficult to determine exactly what were the results of some of the reactions they studied.

The authors made mention of the formation of an acetal in the reaction of diethylzinc with ethanal, but at the same time they expressed their 'surprise' that the acetal apparently contained only one 'ethyl group' instead of the usual two 'ethyl atoms'.

The equation, given in their publication *ad verbatim* reads as follows:

$$2(C_2H_4O + ZnC_2H_5) + H_2O = C_8H_{14}O_2 + C_2H_6 + Zn_2O$$

In none of the reactions reported by these two authors was a product isolated which had resulted from the addition reaction of an organozinc compound to a carbonyl group.

No solvent was used in either of the reactions investigated; diethylzinc was merely warmed with the reactant till it 'had disappeared'.

Frankland; 1863. Frankland [9] was more successful when he investigated the reaction of diethylzinc with ethyl oxalate. He isolated the ester of an acid which had the same molecular formula as 'leucic acid', 2-hydroxyhexanoic acid. Apparently only one of the reactive sites of the oxalic acid ester was attacked by the organometallic reagent:

Zn + O O O O —— 2) $H_3O^{\oplus}$ → HO O O

1.2.4 Barbier Reaction: Saytzeff or Frankland Reaction?

Saytzeff Reaction. In this monograph on the Barbier reaction, it may be of some historical value to trace back the origin of the reaction conditions as published by Barbier at the end of the nineteenth century.

When introducing his arguments for the type of reaction he had investigated, Barbier – in his publication in 1899 [1] – refers to a 'general method developed by Saytzeff' in organozinc chemistry, which he had modified for use with magnesium.

Indeed, even in more modern literature the reaction of an organo halide with zinc in the presence of an organic reactant is sometimes referred to as a 'Saytzeff Reaction' [10–12]. However, after a more accurate study of the chemical literature (see the next Sect. 1.3.2) one is inclined to name this type of reaction after the founder of organometallic chemistry, i.e. Edward Frankland.

Frankland and Duppa; 1865. Two years after his publication of the synthesis of 'leucic acid', Frankland, together with Duppa, reported a crucial improvement of the reaction of diethylzinc with oxalic ester.

Instead of first preparing the ethylzinc compound in advance (using Pébal's [5] or Rieth and Beilstein's [6] methods (vide supra)) the authors introduced the '*in situ procedure*' [13]:

$$2\ C_2H_5I + Zn/Hg + C_2H_5O{-}CO{-}CO{-}OC_2H_5 \xrightarrow[\text{2) } H_3O^{\oplus}]{} HO{-}C(C_2H_5)_2{-}CO{-}OC_2H_5$$

The authors commented:

By this method the necessity of zincethyl is completely by-passed; the whole operation takes place under atmospheric pressure, and the product is obtained in a better yield.

Once again no solvent was used in this reaction: when a 'resinous mass' was formed in the reaction vessel the reaction was considered to be finished and after hydrolysis 'leucic acid ether' was distilled.

This was one of Edward Frankland's last publications in the field of organozinc chemistry; from this time on the pace was set by the school of Alexander Saytzeff in Russia. It was from there that the start of the dominating role of zinc in organometallic intermediates for organic synthesis took place.

A. & M. Saytzeff; 1870. Five years after Frankland and Duppa's publication of the in situ procedure a report from A. Saytzeff appeared in which mention was made of the results, obtained by M. Saytzeff in his (A. Saytzeff's) laboratory [14]: 'Ueber die Wirkung von Zinknatrium auf eine Gemenge von Essigsäure-Anhydrid und Alkoholjodür'.

The following reaction was attempted in order to prepare butanone but the results were not positive:

$$\left.\begin{matrix} CH_3CO \\ CH_3CO \end{matrix}\right\} O + Zn \left\{\begin{matrix} C_2H_5 \\ C_2H_5 \end{matrix}\right. \neq 2 \left.\begin{matrix} CH_3 \\ C_2H_5 \end{matrix}\right\} CO + ZnO$$

Therefore, 'zincethyl' was used in '*Status nascendi*' by adding a 2:1 molar mixture of iodoethane and acetic acid anhydride dropwise to zinc/sodium (no mention was made of a solvent). After 20 to 30 h standing at room temperature the contents of the flask were carefully hydrolyzed and, finally, distilled. The product obtained was butanone.

In the same way acetone could be prepared by using iodomethane as the haloalkane.

In a final remark M. Saytzeff announced his plan to prepare 3-methyl-2-butanone with the aid of 2-halopropane. However, another five years passed before further reports on organozinc chemistry appeared from A. Saytzeff's laboratory.

1.2.5 Zinc in Organic Chemistry

Saytzeff's Contributions – Wagner, Kanonnikov, M. Saytzeff; 1875. At A. Saytzeff's laboratory, Wagner [15] repeated the unsuccessful reaction of ethylzinc with ethanal as published in 1863 [8] by Rieth and Beilstein (vide supra). 2-Butanol was obtained in good yield:

Zn + 2 O H → 2) $H_3O^{\oplus}$ OH

In the same year, three further publications from Saytzeff's laboratory drew special attention.

Kanonnikov [16] synthesized 2-butanol via an unusual procedure: in a one-step, Frankland-type, reaction in which two different haloalkanes were allowed to react with zinc in the presence of ethyl methanoate:

I + I + 2 Zn + H O O → 2) $H_3O^{\oplus}$ OH

The difference in reactivity of the two iodoalkanes causes this reaction to proceed in two consecutive steps (see also [17]).

Wagner [18] reported the synthesis of 3-pentanol by the same type of procedure:

2 I + 2 Zn + H O O → 2) $H_3O^{\oplus}$ OH

Finally, M. Saytzeff [19] published his results of the same type of reaction with allyl iodide:

$$2\ \ldots I + 2\ Zn + H\ldots O \ldots \xrightarrow[2)\ H_3O^{\oplus}]{} \ldots OH$$

This was the start of a series of publications on one-step reactions with allylic halides – the Jaworsky reaction – which will be further discussed in Chapter Two.

Wagner; 1877 and 1881. Further publications from Wagner [20] and [21] using acrolein, butanal and benzaldehyde as the carbonyl compounds – performed in the classical two-step fashion – showed the great versatility of the organozinc reagent in organic synthesis:

$$\ldots Zn \ldots + 2\ \ldots H \xrightarrow[2)\ H_3O^{\oplus}]{} 2\ \ldots OH$$

$$\ldots Zn \ldots + 2\ \ldots H \xrightarrow[2)\ H_3O^{\oplus}]{} 2\ \ldots OH$$

$$\ldots Zn \ldots + 2\ \ldots H \xrightarrow[2)\ H_3O^{\oplus}]{} 2\ \ldots OH$$

It would be true to say that by the late 1880s organozinc reagents had built up a reputation for the synthesis of alcohols.

Reformatsky; 1887. Finally, as a culmination of the development of the zinc-tradition, Reformatsky [22], again at Saytzeff's laboratory, introduced the well-known one-step reaction of 2-haloalkanoic acid esters with zinc in the presence of a carbonyl compound, which leads to the formation of 3-hydroxyalkanoic acid esters.

$$X \ldots O \ldots + Zn + \ldots \xrightarrow[2)\ H_3O^{\oplus}]{} HO \ldots O \ldots$$

1.3 The Early History of Organomagnesium Chemistry

1.3.1 Pre-Barbier period

Hallwachs and Schafarik; 1859. Cahours; 1860. Ten years after Frankland's discovery of the organozinc compounds Hallwachs and Schafarik [23] and then Cahours [24], were able to prepare and identify dialkylmagnesium compounds.

Löhr; 1889. Fleck; 1893. However it was a further thirty years before Löhr [25] and Fleck [26] started investigating the reactions of such compounds.

Ironically, in an attempt to prepare phenylmagnesium bromide which was unknown at that time, Fleck added bromine to a suspension of diphenylmagnesium in diethyl ether[1]. From the equations given in his publication it can be concluded that he should have foreseen the formation of bromobenzene and magnesium bromide on the addition of bromine:

$$(C_6H_5)_2Mg + 2Br_2 \longrightarrow 2\,C_6H_5\text{-}Br + MgBr_2$$

Nevertheless Fleck added bromine until the reaction mixture had attained a yellowish color:

Fleck reports that diethyl ether, together with the 'kleinen Ueberschuss an Brom', (i.e. the slight excess of bromine) were evaporated after the reaction was finished and that he failed to isolate anything other than bromobenzene.

One can only speculate how organomagnesium chemistry would have had developed if this young chemist had not made this error (which incidentally, had not been noticed by the illustrous Lothar Meyer at whose laboratory Fleck did his research, nor by the referees of the Annalen [27], the journal in which publication took place).

1.3.2 Magnesium in Synthetic, Organic Chemistry

Barbier; 1899. A few years later Barbier investigated using magnesium in place of zinc in what he named *la méthode générale imaginée par Saytzeff*, the Saytzeff reaction, for the preparation of a tertiary alcohol, using diethyl ether as the solvent [1]:

$$\text{ketone} + Mg + CH_3I \xrightarrow[\text{2) } H_3O^{\oplus}]{Et_2O} \text{tertiary alcohol (OH)}$$

---

[1] One wonders about the quality of this diphenylmagnesium prepared by Fleck, since the compound is known to be readily soluble in diethyl ether.

Grignard; 1900. However, it was Barbier's student Victor Grignard [28] who in fact was responsible for the introduction of magnesium for synthetic purposes in organic chemistry by using the stepwise procedure in which first an organomagnesium compound – now widely known as a 'Grignard reagent' – is prepared, after which this is brought into interaction with a selected substrate without it actually being isolated and/or purified.

Name of the One-step Procedure. Early publications in this new field of organic chemistry reveal a bitter rivalry over priority and eponimy between the student/assistant Grignard (Nobel-prize winner 1912) and the teacher/head-of-the-department Barbier. Several modern historians have come to the conclusion [29] that the most reasonable name for this step-wise procedure would have been Barbier-Grignard reaction.

The names, given to the one-step procedure, as first published by Barbier, vary from country to country and from individual to individual.

The most systematic name would be the one in which the procedure is made clear to the reader such as *one-pot reaction* (e.g. in [30] and [31]) or *one-flask reaction, in situ reaction* [32], *one-step reaction* [33], etc. and such names are often encountered. However more often names such as *(modified) Wagner-Saytzeff reaction, Saytzeff reaction, Barbier-Grignard reaction* [29], [34], [35] and [36], *batch-method* [36], *Dreyfuss-Barbier reaction* [37], *Jaworsky reaction* [38], *Grignard-type reaction* [39], [40], etc. are used.

Probably because of unfamiliarity with his early work the name *Frankland reaction* has not been used although, as has been made clear earlier, many good reasons exist for such a claim.

More recently [41–43] the name *Barbier reaction* has also been linked up to one-step processes in which metals, different from magnesium, are applied (see chap 3). In this monograph preference is given to names for such processes in which the applied metal is recognizable; hence names such as Li-Barbier or Zn-Barbier etc. reactions will be used.

1.4 The Barbier Reaction

Unknown Reaction. It is of historical and scientific interest to consider the reasons why the Barbier reaction has only recently drawn the significant attention of synthetic organic chemists. The first review article of the field was published in 1977 [44], more than three quarters of a century after the reaction was first discovered.

Thus it appears that several publications from the first years of application of the Grignard reagent, in which the versatility of the one-step Barbier reaction was well demonstrated, did not meet with the response they merited.

In particular the following two reports in 1903 and 1911 clearly indicated the advantages of this simple and quick procedure.

Houben; 1903. Houben [45] showed the effectiveness of the Barbier reaction in his synthesis of 3-butenoic acid:

$$\text{CH}_2{=}\text{CHCH}_2\text{Br} + \text{Mg} + \text{CO}_2 \xrightarrow[2)\ \text{H}_3\text{O}^{\oplus}]{} \text{CH}_2{=}\text{CHCH}_2\text{CO}_2\text{H}$$

and this publication was among the first in a series of reports on the use of the one-step procedure with magnesium as the metal. In particular from Reformatsky's laboratory at Kiev soon after 1903, the successful preparation of homoallylic secondary and tertiary alcohols using this procedure started (see Sect. 2.3).

Davies and Kipping; 1911. The strongest support for the use of the Barbier procedure came from Davies and Kipping [46] who concluded their publication on the synthesis of alcohols with the aid of organomagnesium compounds, with the following statement:

The results have shown that in these reactions, also, the preliminary preparation of the magnesium compound is unnecessary, and that a good yield of the desired product is obtained by gradually adding a mixture of the alkyl or aryl halogen compound with the aldehyde, ketone or ester, to the theoretical quantity of magnesium, which is covered with a little ether.

A little further on these two authors continue:

.. but it is hoped that the publication of this note will save time and trouble to many who are working with Grignard reagents.

It is strange that the use of the Barbier reaction was virtually abandoned soon after the publication of the Grignard reaction, though this reaction is often more time-consuming and usually no more successful.

Even for reactions where the Barbier procedure has clearly and unambiguously proved to be superior over the Grignard-type of reaction, this two-step procedure was, till recently, much more generally used.

A tentative explanation of this phenomenon will be given in some of the following chapters.

1.5 More Recent Developments in Barbier Chemistry

Reintroduction of Zinc. Perhaps one of the most striking developments in modern Barbier chemistry is the increasingly successful reintroduction of zinc as the metal for this category of condensation reactions. In fact, zinc had never completely lost favour in organic synthesis due to its application in the Reformatsky reaction (Sect. 1.2.5).

In some instances magnesium is preferred for this reaction when a need exists for a greater reactivity of the intermediate organometallic derivative. On the other hand such a greater reactivity may cause the formation of undesired sideproducts, due to unforeseen reactions. See also Sect. 2.2.

The general tendency for the application of zinc in Barbier-type reactions has to be attributed to the decreased reactivity of the zinc-derivatives: organometallic reagents of reduced reactivity have become of growing importance as modern organic chemistry tends to the synthesis of increasingly delicate targets.

New Metals; New Techniques. It has been made clear in this introductory chapter that the Barbier reaction was a new branch of synthetic chemistry, developed from the organozinc chemistry that enjoyed great success at the end of the nineteenth century. Zinc was favoured at that time, probably because of its greater availability and, no doubt, because the organozinc compounds were easier to isolate and handle.

However, this reduced reactivity, for example in reactions with ketones and also when aromatic halides are used, of course also had its disadvantages. The introduction of magnesium opened up new ways to numerous organic molecules and the explosive use of organomagnesium compounds, Grignard reagents, reflected the great need felt by organic chemists for more effective synthetic tools. Still more reactive organometallic reagents were introduced afterwards in the beginning of this century of which the organolithium and -aluminium compounds need special mention.

Developments of new working techniques such as better glassware, purified inert gasses, low temperature techniques, new solvents and many others have introduced new synthetic procedures which were simply inconceivable in the Saytzeff and Barbier-Grignard period.

1.6 Contents of the Following Chapters

Chapter 2 of this monograph will present as complete a review as necessary of the application of the Barbier reaction procedures as published till 1990 and where possible till 1991.

Chapter 3 will deal with the application of other metals in Barbier-type reactions, quite often also referred to as Saytzeff reactions, Grignard reactions, etc.

Chapter 4 will deal with the mechanistic aspects of the heterogeneous reactions in which metals and organic halides are involved.

It was decided to give only a limited amount of information on the Reformatsky reaction (see Sect. 2.2) and not to deal at all with the Simmons-Smith reaction in which an intermediate carbenoid zinc reagent is involved [47]:

I⌒I + Zn ⟶ [I⌒ZnI] ⟶

Chapter 5, finally, will give detailed information on the different technical aspects for the successful performance of Barbier-type reactions.

1.7 References

1. Barbier Ph (1899) Compt Rend 128: 110
2a. Frankland E (1849) Ann 71: 171
2b. Frankland E (1849) J Chem Soc 2: 263
3a. Frankland E (1849) Ann 71: 213
3b. Frankland E (1849) J Chem Soc 2: 297
4. Frankland E (1855) Ann 95: 28
5a. Pébal L (1861) Ann 118: 22
5b. Pébal L (1862) Ann 121: 105
6. Rieth R and Beilstein F (1862) Ann 123: 245
7. Freund A (1861) Ann 118: 1
8. Rieth R and Beilstein F (1863) Ann 126: 241
9. Frankland E (1863) Ann 126: 109
10. Jaworsky W (1909) Ber 42: 435
11. Miginiac-Groizeleau L (1961) Ann Chim (Paris) 13: 1071
12. Joffe ST, Nesmeyanov AN (1967) The Organic Compounds of Magnesium, Beryllium, Calcium, Strontium and Barium. North Holland, Amsterdam
13. Frankland E and Duppa BF (1865) Ann 135: 25
14. Saytzeff A (1870) Z. f. Chemie 13: Neue Folge 6, 104
15. Wagner E (1875) Bull Soc Chim France 25: 289
16. Kanonnikoff J and Saytzeff A (1875) Ann 175: 374
17. Nützel K (1973) In Methoden der Organischen Chemie (Houben-Weyl) Vol 13/2a Metallorganische Verbindungen. Georg Thieme, Stuttgart, p 552
18. Wagner G and Saytzeff A (1875) Ann 175: 351
19. Saytzeff M (1875) Bull Soc Chim France 25: 297
20. Wagner G (1877) Ber 10: 714
21. Wagner G (1881) Ber 14: 2557
22. Reformatsky A (1887) Ber 20: 1210
23. Hallwachs W and Schafarik A (1859) Ann 109: 201
24. Cahours A (1860) Ann 114: 240
25. Löhr Ph (1891) Ann 261: 72
26. Fleck H (1893) Ann 276: 129
27. Gilman H, Chem and Eng News, 1977: March 28, p. 49.
28. Grignard V (1900) Compt Rend 130: 1322
29. Urbansky T (1978) Chem in Britain 12: 191
30. Kitazume T and Ishikawa N, Chem Lett 1981: 1679
31. Mandai T, Nokami J, Yano T, Yoshinaga Y and Otera J (1984) J Org Chem, 49: 172
32. Tanaka H, Hamatani T, Yamashita S and Torii S, Chem Lett 1986: 1461
33. Molnarfi S (1964) Brevet d'Invention No. 1359, 195; Chem Abstr 61: P8236g (1964)
34. Obe Y, Sato M and Matsuda T, (1972) Kyushu Daiyaku Kogaku Shuhu 1971: 208; Chem Abstr 77: 48546z
35. Katzenellenbogen JA and Lenox RS (1973) J Org Chem 38: 326
36. Henze HR, Alan BB and Leslie WB (1942) J Org Chem 7: 326
37. Richet G and Pecque M (1974) Compt Rend, 278: 1519
38. Jaworsky W, J Russ (1908) Phys-Chem Soc 40: 782; ibid 1746 Chem Zentralblatt, 1908 II, 1412; ibid. 1909 I, 856
39. Mykaiyama T and Harada T, Chem Lett 1981: 1527

40. Wada M and Akiba K (1985) Tetrahedron Lett 26: 4211
41. Molle G and Bauer P (1982) J Amer Chem Soc 104: 3481
42. Shih N-Y, Mangiaracina P, Green MJ and Ganguly AK (1989) Tetrahedron Lett 30: 5563
43. Luche J-L (1990) In: Mason TJ (Ed) Advances in Sonochemistry, Vol 1 JAI Press, Greenwich, p 126
44. Blomberg C and Hartog FA, Synthesis 1977: 18
45. Houben J (1903) Ber 36: 2897
46. Davies H and Kipping FJ (1911) J Chem Soc 99: 296
47a. Simmons HE and Smith RD (1958) J Amer Chem Soc 80: 5323
47b. Simmons HE and Smith RD (1959) J Amer Chem Soc 81: 4256
47c. Furukawa J and Kawabata N (1974) Adv. Organomet Chem 12: 83

2 Synthetic Applications of the Barbier Reaction

2.1 Synthetic Procedures: One-Step or Two-step?

Synthetic Guideline. In general, the synthetic organic chemist, when planning the synthesis of an organic compound, is often guided by the experience of others. Previous reports of successes or failures of certain types of reactions, can stimulate or discourage the researcher to apply them in order to achieve the goal that was set.

In this monograph one aim is to provide guidelines for the choice-making procedure when organometallic reagents are to be employed for synthetic purposes. Roughly speaking in this case the choice is: should we make use of a one-step or of a two-step procedure?

Why One-Step? Two questions need to be answered:

a. What made Frankland in 1865 [1] decide to use the one-step synthesis of the 'leucic acid' by reacting iodoethane with zinc and methyl oxalate more than thirty years before Barbier's experiment with magnesium as the metal (Sect. (1.2.4)?

2 I + 2 Zn (Hg) + O O O O → 2) $H_3O^{\oplus}$ HO O O

b. What made Grignard in 1900 [2] decide to follow a two-step procedure for the synthesis of an alcohol after his supervisor Barbier [3] had successfully prepared dimethylheptenol in one step one year earlier (Sect. 1.3.2)?

For the 'Frankland reaction' the reasoning behind the choice was rather obvious: it avoided the tedious and risky reaction-step to prepare ethylzinc.

The same must have been true of the researchers that continued this type of work at Saytzeff's institute (Sect. 1.2.5). The synthesis of alcohols and of β-hydroxyacid esters was considerably simplified by avoiding the preliminary preparation of the organozinc compound.

On the other hand it may just have been scientific interest or perhaps a spark of inspiration that led Grignard to perform the Barbier synthesis in two consecutive steps; allowing the chemist to first prepare, and if necessary, isolate and purify, the reactive intermediate before continuing the procedure.

The tremendous success of the two-step synthetic procedure leaves no doubt that Grignard's choice was correct:

In order to keep matters well-ordered and controllable, do not do in one step what can be done in two.

The forthcoming information in this chapter reports the deviation from Grignard's 'advice' by others after him, discussing both the reasons why this was done as well as the advantages and disadvantages of their choice.

2.2 The Replacement of Zinc by Magnesium in the Reformatsky Reaction

It was from Saytzeff's laboratory (Sect. 1.2.5) that Reformatsky [4] first reported the successful synthesis of 3-hydroxyalkanoic acid esters with the aid of an intermediate organozinc compound, thus developing the reaction which still bears his name:

X, O, O + Zn + O —— 2) $H_3O^{\oplus}$ → HO, O, O

This reaction is one of the important synthetic tools, left over from the zinc-dominated organometallic chemistry at the end of the nineteenth century.

As would be expected, very shortly after Barbier's and Grignard's discovery of the effective replacement of zinc by magnesium in the synthesis of alcohols, publications appeared regarding the use of magnesium in the Reformatsky reaction.

From the rather large number of reports in this field the following typical ones have been selected for discussion in this chapter.

As early as 1901, two years after Barbier's introduction of magnesium in organic synthesis, the successful use of magnesium in the Reformatsky reaction was reported [5].

3-Methylcyclohexanone as well as cycloheptanone were reacted with ethyl 2-bromo-2-methylpropanoate and magnesium in diethyl ether. Yields were as high as 50% which was remarkable as a previous attempt in that same year [6], to react that same halo-ester with zinc in the presence of 3-methylcyclohexanone had failed, most likely due to steric hindrance:

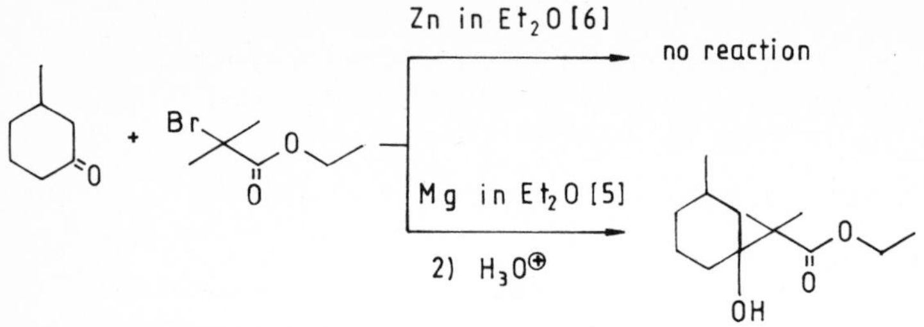

Zn in Et_2O [6]

no reaction

Mg in Et_2O [5]

2) $H_3O^{\oplus}$

OH

A few years later superior results were also reported when using magnesium in the Reformatsky reaction with acetophenone and ethyl 2-haloethanoate, with benzene as the solvent [7].

Equimolecular mixtures of both reactants were dissolved in benzene, the reaction was started by adding a few chips of the metal and the reaction mixture was kept boiling until the equivalent amount of magnesium chips had been added.

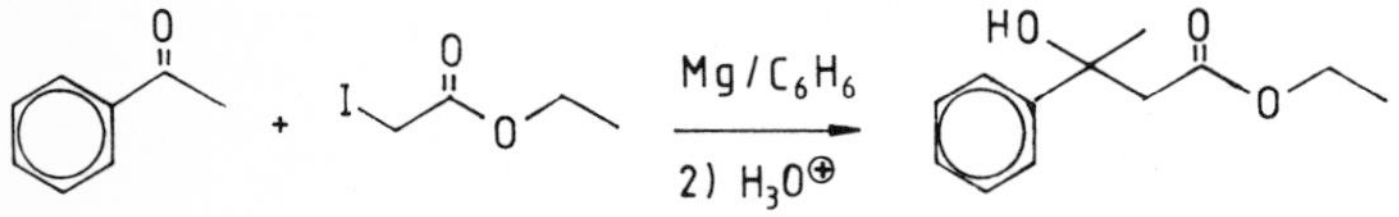

Mg/C_6H_6

2) $H_3O^{\oplus}$

In a subsequent publication [8] results of this type of reaction with 4-methylacetophenone, propiophenone, butyrophenone, isovalerophenone and capronophenone were reported. In some instances the reaction proceeded violently.

Claims were refuted [9] that in Reformatsky reactions with ethyl bromoethanoate and acetophenone (or benzophenone), magnesium gave results inferior to those obtained with zinc [10].

Furthermore new results were mentioned [9] of reactions with 4-methoxy- as well as with 4-ethoxyacetophenone, with butyrophenone and with valerophenone; each gave the expected products in good yields.

In general, the greater reactivity of the organomagnesium intermediate in the modified Reformatsky reaction allows it to be utilized for sterically hindered reactions which would not occur with zinc.

For instance, in 1919, tetramethylhexanoic acid was prepared [11] in a multi-step synthesis which started with the following modified Reformatsky reaction:

+ Mg

2) $H_3O^{\oplus}$

OH

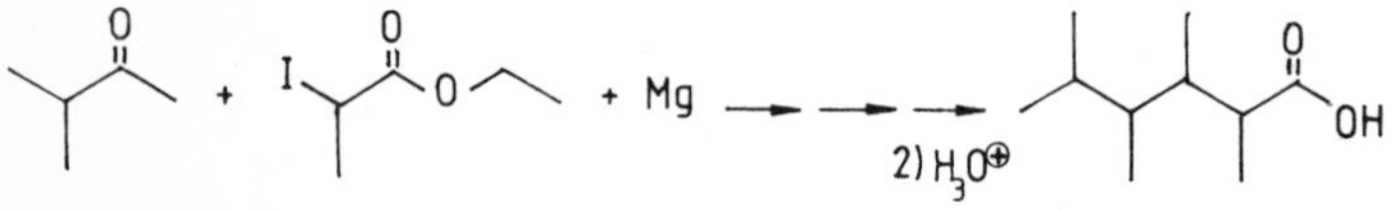

Much more recently a comparison between the results obtained from Reformatsky reactions with zinc and magnesium in which sterically hindered α-bromo esters were used was published [12].

The results, presented in Table 2.1, clearly indicate the advantage of the use of magnesium.

When the zinc is replaced by the more reactive magnesium, the halomagnesium enolate attacks also the carbonyl carbon of the ester to form β-ketoesters. In order to retard this possible self-condensation it was recommended to use the more bulky *tert*-butyl esters of the α-haloalkanoic acids [13]:

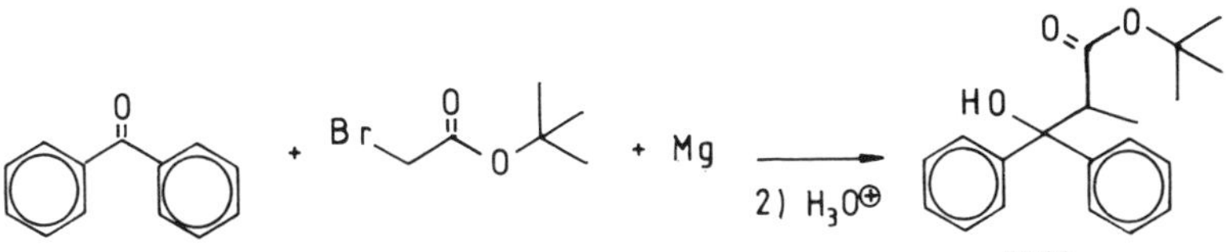

+ Mg, 2) $H_3O^{\oplus}$ → 81 %

Table 2.1. Results of 'Zn- and Mg-Reformatsky reactions' with sterically hindered reagents [12]

α-Halo ester	Carbonyl compound	Yields obtained with Zn	Yields obtained with Mg
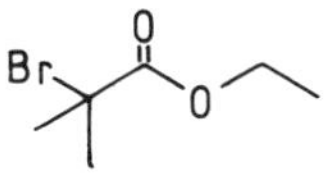	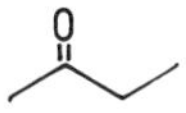	55%	91%
idem	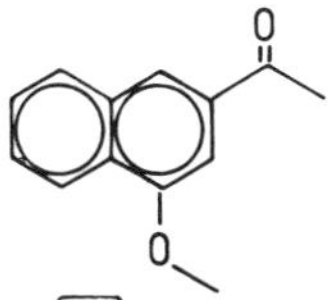	16%	56%
idem	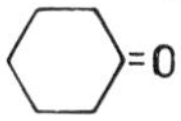	6%	19%
		-	-
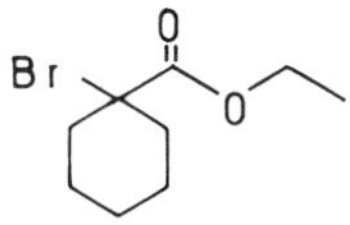	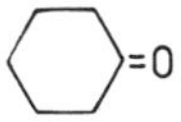	20%	88%
idem	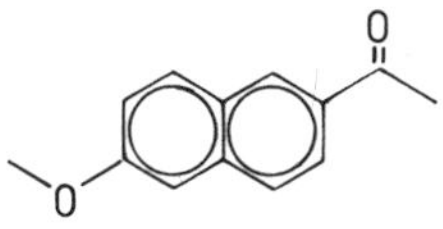	9%	62%

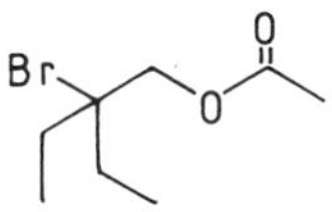

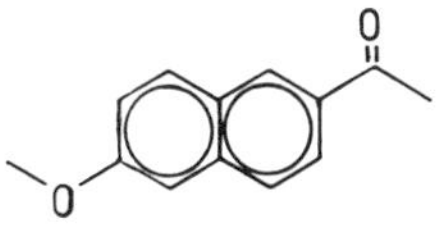

Earlier no reaction was observed in the equivalent procedure with the ethyl bromopropanoate [14]:

+ Zn → 2) $H_3O^{\oplus}$ no products

Other ketones, used to investigate the reactivity in this modified Reformatsky reaction with *tert*-butyl bromoethanoate were acetone (69% yield), cyclohexanone (80% yield), 3-methyl-2-butanone (78% yield), benzophenone (74% yield), 4-phenyl-4-phenylbut-3-en-2-one (benzal-acetone; 42% yield) and 1,4-diphenylbut-3-en-2-one (40% yield). The conclusion was made by the author that the

present method is more convenient than the procedure using zinc.

It can be said therefore that some remarkable results have been reported when magnesium was used in Reformatsky reactions. In general, however, the use of this metal leads to undesired products due to the high reactivity of the intermediate organomagnesium compound as demonstrated in the following two examples:

A direct displacement of the ester alkoxy-group may take place with magnesium, resulting in the formation of a β-keto-acid as was discovered rather early in the history of organomagnesium chemistry [15]:

Mg, 2) $H_3O^{\oplus}$

67%

The same type of products was also obtained with other esters such as those of 2-bromopropanoic acid, 2-bromobutanoic acid and 2-bromo-3-methylbutanoic acid.

In that same period, identical problems were encountered when ethyl oxalate was used in Reformatsky reactions [16]: with zinc the expected hydroxy acid ester was obtained whereas magnesium yielded the keto acid ester.

Zn 2) $H_3O^{\oplus}$

Mg 2) $H_3O^{\oplus}$

Neither Shriner [17] in 1942, nor Rathke [18] in 1974, in their review articles on the Reformatsky reaction have come to the conclusion that zinc could in general be replaced by magnesium in this synthetic procedure.

However, as has been demonstrated above, in special instances magnesium is to be preferred.

2.3 Barbier Reactions with Allylic Halides

2.3.1 Introduction

As mentioned in Sect. 1.2.4, Alexander Saytzeff and his coworkers, at the end of the nineteenth century, started using allylic halides in the 'one-batch', 'one-pot' or 'one-step' synthesis of alcohols, a procedure often referred to as the Saytzeff reaction, e.g. by Barbier [3].

In his doctoral thesis Grignard had come to the conclusion that 'his reaction' could not be applied to allylic halides [19]. Attempts to prepare allylmagnesium halides led to unidentifiable products and the results of their reactions with e.g. aldehydes were

in general mediocre and inferior to those obtained by the use of zinc in Saytzeff's method.

Houben was among the first to adapt Barbier's synthetic procedure. Just as Grignard [19] and Arbusov [20] (see [21]) had experienced before him, he was unsuccessful in his attempts to prepare allylmagnesium halides which he required for the synthesis of 3-butenoic acid [22].

Some success was met however in the one-step 'Saytzeff reaction' in which magnesium was reacted with allyl bromide under an atmosphere of dry carbon dioxide:

Br + CO_2 + Mg → (2) $H_3O^{\oplus}$) OH, O

11%

Only 19% of the magnesium remained unreacted and, after repeated distillation, Houben isolated the unsaturated acid in 11% yield.

In an earlier publication [21] it had been reported, that, on heating allyl iodide with magnesium and acetophenone, Arbusov had observed a violent reaction but none of the expected homoallylic alcohol could be isolated.

However, replacing magnesium with zinc (!!!), and using diethyl ether as the solvent (a procedure he referred to as the M. and A. Saytzeff method) allowed Arbusov [20] to obtain the desired product in a 31.5% yield;

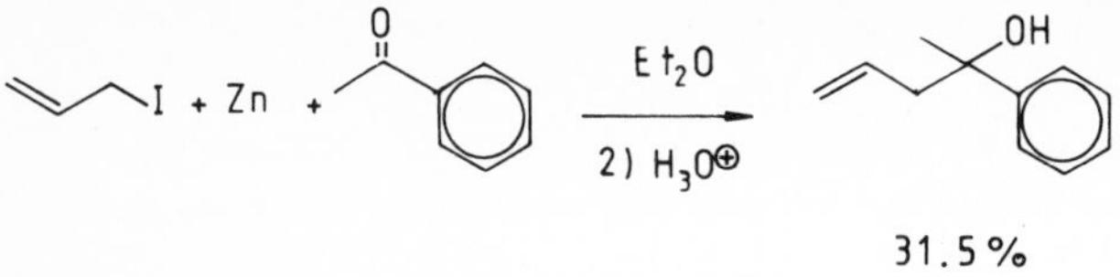

Et_2O

2) $H_3O^{\oplus}$

31.5%

Another noteworthy, early, publication mentions the synthesis of 3-buten-1-ol by a Barbier reaction of allyl bromide, magnesium and trioxymethylene [23] (see also Sect. 2.4).

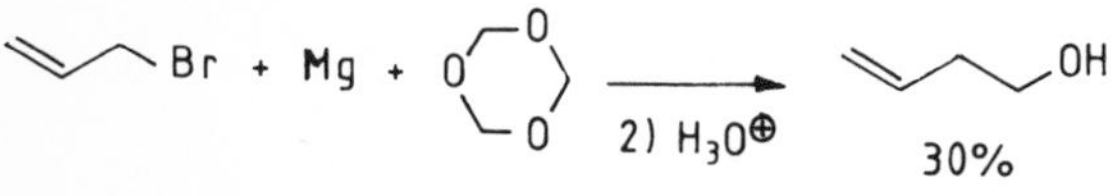

2) $H_3O^{\oplus}$

30%

2.3.2 The Jaworsky Reaction

It was through Jaworsky's work at Reformatsky's laboratory, that the use of the one-step Barbier reaction for the preparation of homoallylic alcohols became a reaction of fundamental importance in organic synthesis [21]. It is therefore of no surprise that this procedure is often referred to as the 'Jaworsky reaction'.

Jaworsky, followed by many other researchers in and outside Reformatsky's research group, applied the one-step 'Saytzeff-Wagner' method in which zinc was replaced by magnesium:

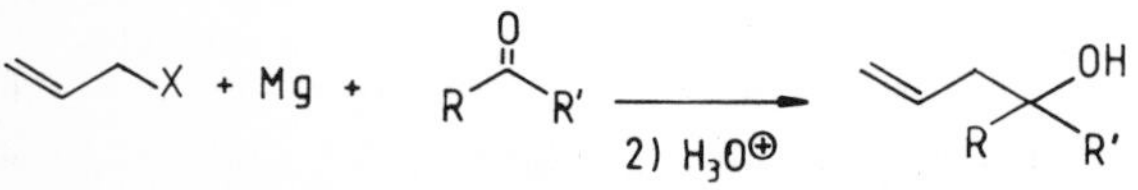

2) $H_3O^{\oplus}$

Table 2.2 presents some of the results obtained with the Jaworsky reaction in the beginning of this century and demonstrates its versatility as well as typical yields.

One of the possible dangers of a one-pot execution of a reaction is that it leads to the simultaneous presence in the reaction mixture of the products and the reactants; this can cause uncontrolled reactions between these reagents.

A very obvious one in this context would be the Williamson-type ether-formation reaction which indeed was observed during a Jaworsky reaction which included allyl chloride and propanone earlier in this century [40]:

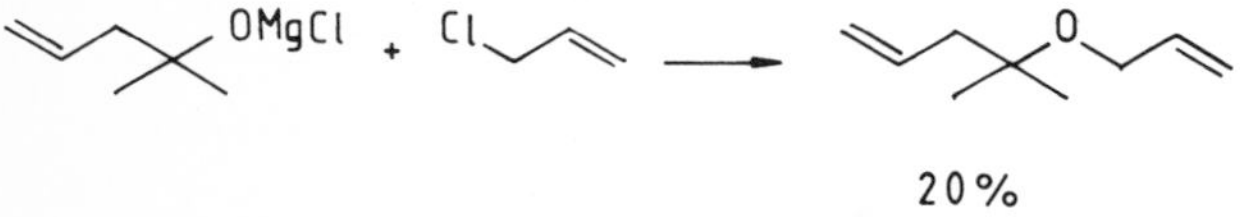

20%

Table 2.2. Results of some Jaworsky reactions as published at the beginning of this century.

Carbonyl Compound	Product	Yield	Ref.
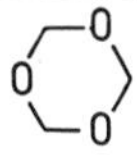 O O O	 OH	30%	23
$(CH_2O)_n$	 OH	15-20%	32
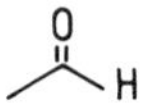 O H	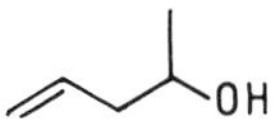 OH	11-20%	34
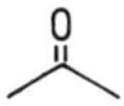 O	 OH	75%	21
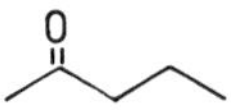 O	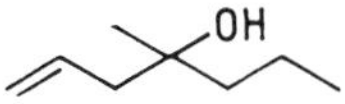 OH	88%	21
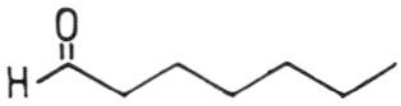 H O	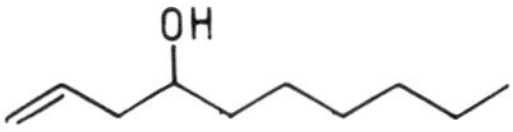 OH	48%	21
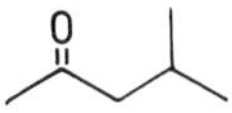 O	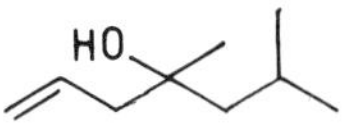 HO	40-60%	31
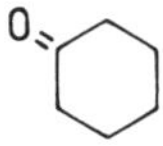 O	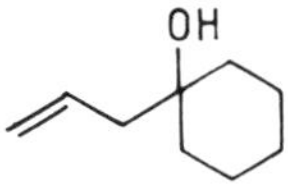 OH	81%	29
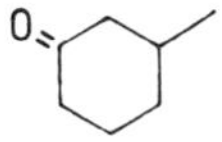 O	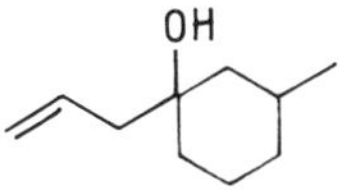 OH	98,7%	29
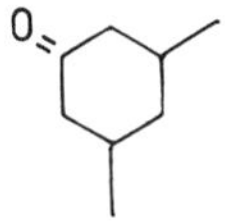 O	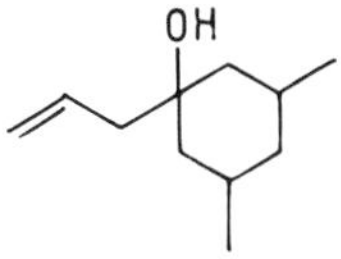 OH	95%	29
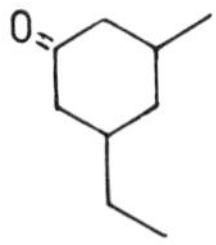 O	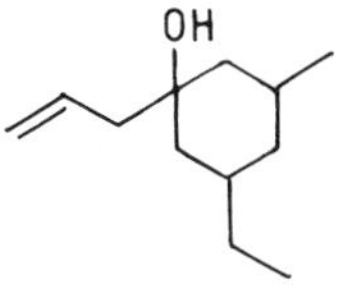 OH	99.2%	29
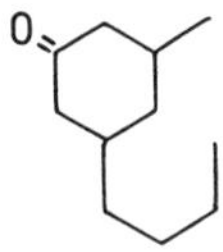 O	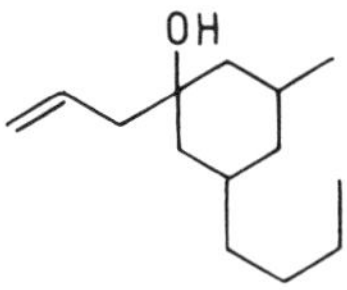 OH	97.1%	29

Table 2.2 (continued)

Carbonyl Compound	Product	Yield	Ref.
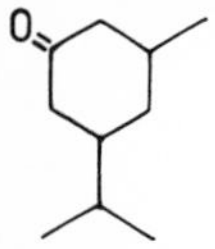	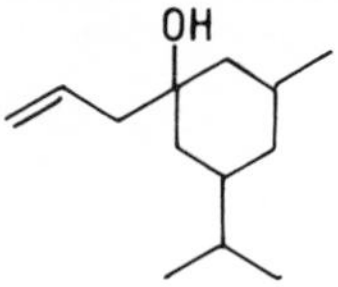	91,2%	29
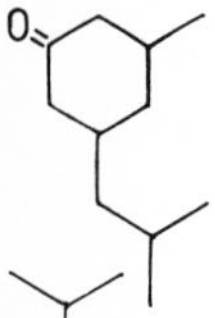	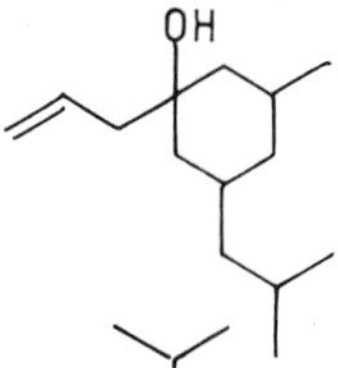	98%	29
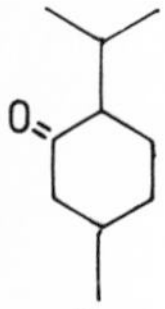	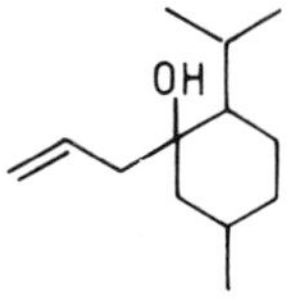	97%	21,39
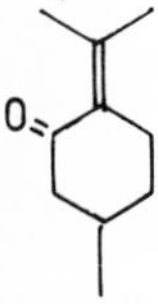	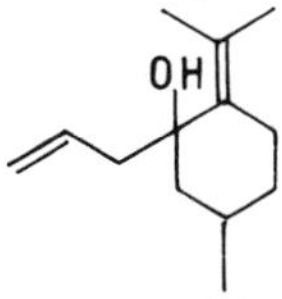	90%	21
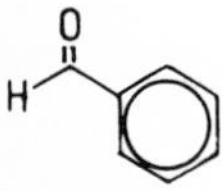	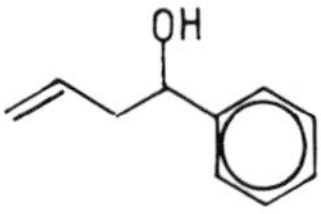	-	34
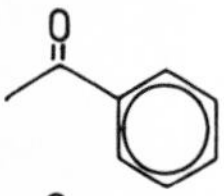	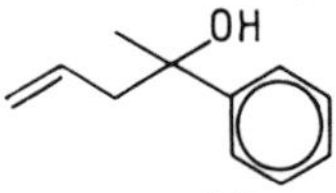	90%	21
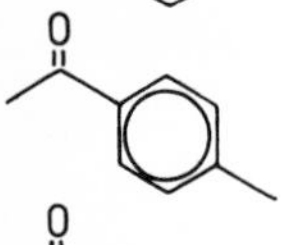	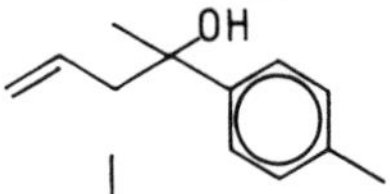	90%	21,24
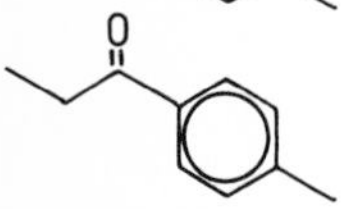	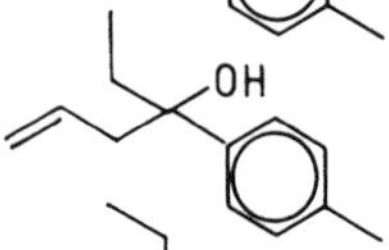	75%	24
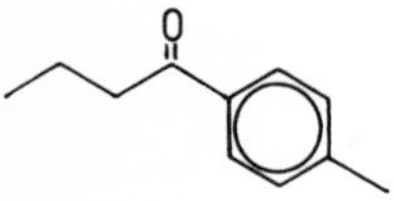	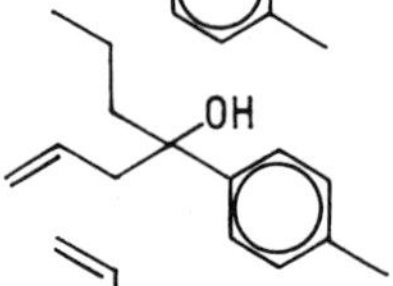	75%	24
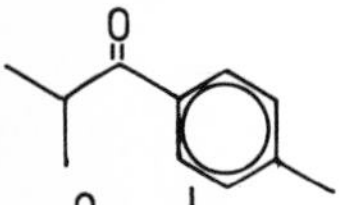	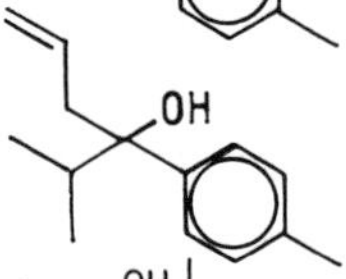	-	24
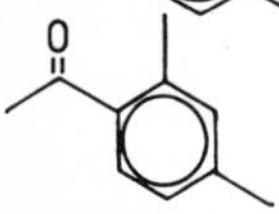	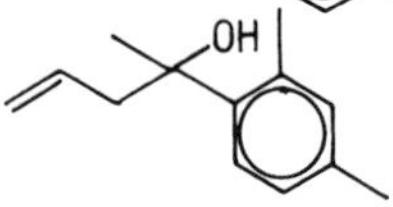	100%	29

Table 2.2 (continued)

Carbonyl Compound	Product	Yield	Ref.
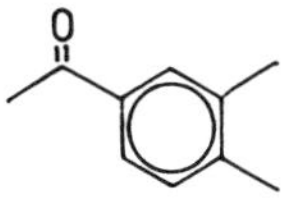	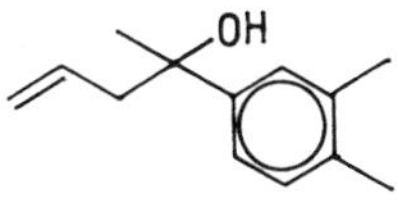	96%	29
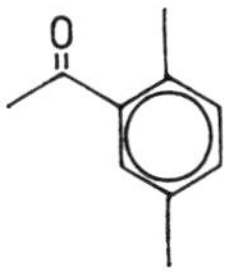	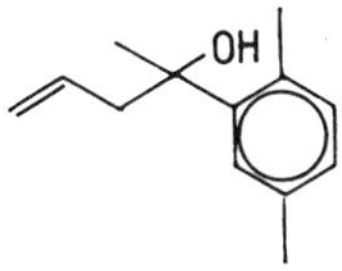	100%	29
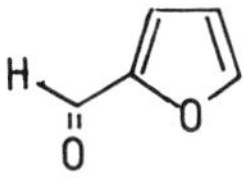	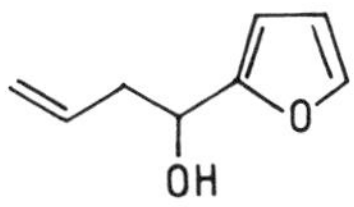	78%	28
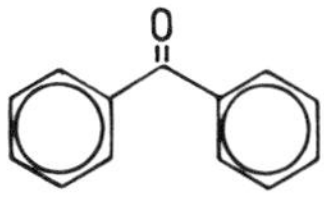	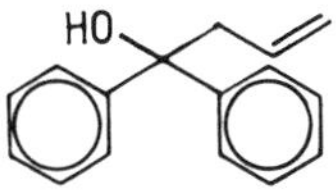	60%	21, 25
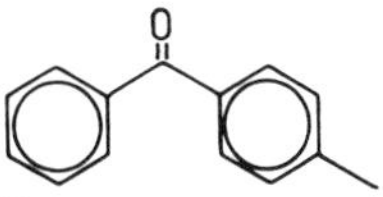	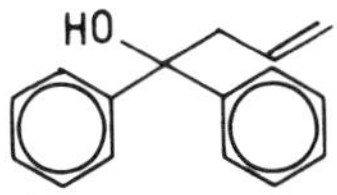	90%	26
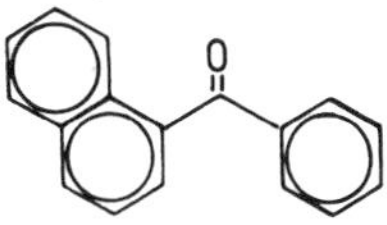	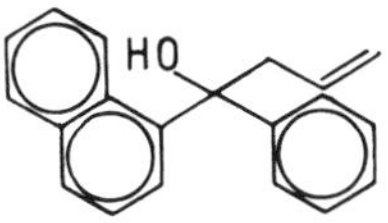	-	35
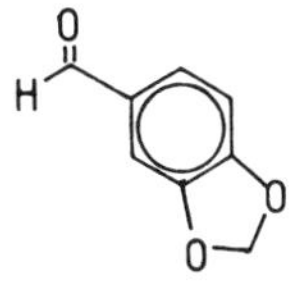	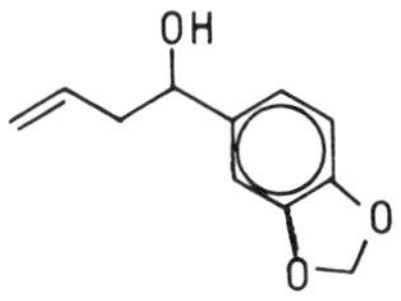	-	36
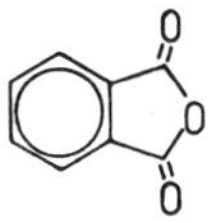	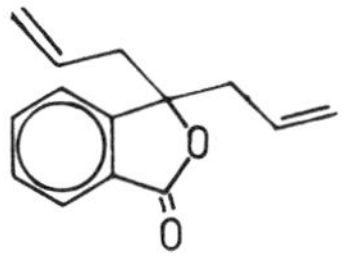	75%	37
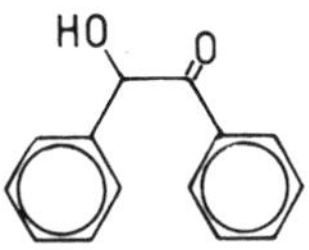	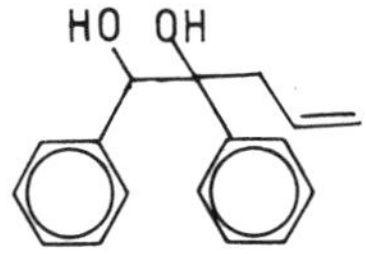	90%	38
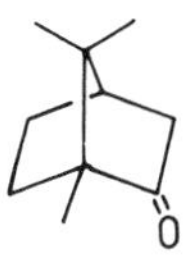	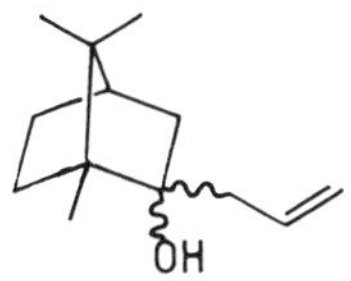	96%	21, 30

It should be mentioned here however that products of this type are seldom reported and therefore seem not to be formed in considerable quantities. The reaction of the halide with the metal and consecutive reaction of the intermediate organometallic species (see Sect. 4.4 for a discussion of the mechanism of the Barbier reaction) with the carbonyl compound seems to be fast enough to prevent this ether formation.

2.3.3 Allylmagnesium Halides

An interesting event in the history of the Barbier reaction was the publication in 1928, by Gilman and McGlumphy, of a new procedure to prepare Grignard reagents from allylic halides [41].

These authors demonstrated that, under carefully controlled reaction conditions, stable allylic-magnesium halides could be formed. The Wurtz-type coupling reaction which leads to the formation of 1,5-hexadiene and which is responsible for the loss of most, if not all, of the allylic halide could be suppressed almost completely:

(Wurtz-type reaction)
(+ X)
X + Mg
MgX (Grignard reaction)

A comparison was made between the results of reactions of this allylic Grignard reagent and those of one-step Barbier reactions of allyl bromide and magnesium with three different substrates, i.e. benzophenone, acetophenone and carbon dioxide.

These results (see Table 2.3) seem to indicate that the two-step procedure is to be preferred to the one-step Barbier reaction.

In all fairness, however, it should be noted that only yields were quoted from previous reactions obtained in the very beginning of the development of Barbier chemistry and that no reexamination had taken place of these reactions under comparable conditions.

More recent results from systematic investigations of the Barbier reaction clearly speak in favour of the one-step reaction of allylic halides as will be demonstrated in Sect. 2.3.4. Furthermore chap 3 will show that such one-step reactions can also give excellent yields with metals other than magnesium.

Table 2.3. Comparison of the yields of reactions of allyl bromide and magnesium with benzophenone, acetophenone or carbon dioxide in the one-step and two-step reaction [41]

Product	Yield in % via RMgBr	Yield in % via One-step reaction	Ref.
[structure: 1,1-diphenylbut-3-en-1-ol]	75	60	21, 25, 26
[structure: 2-phenylpent-4-en-2-ol]	80	60	21
[structure: but-3-enoic acid]	22	11	22

2.3.4 More Barbier-Type Reactions with Allylic Halides

Knorr; 1932. The application of Gilman and McGlumphy's method for the preparation of allylic Grignard reagents does not look very attractive for industrial use; only two years after its publication, the first patent for the preparation of homoallylic alcohols via the one-step Barbier reaction was applied for [42].

For this reaction the use was recommended of a hydrocarbon solvent (such as benzene, xylene, decahydronaphthalene, cyclohexane or a petroleum ether (with preference for an aromatic hydrocarbon)) mixed with diethyl ether. Thus the dangers of large-scale use of the extremely risky ether were largely suppressed.

Table 2.4 presents the results mentioned in the patent.

It is to be noted that besides allyl bromide, reactions with 3-methylallyl bromide are mentioned in this patent. What was unknown to Knorr at that time is that the products from the organometallic reagent of this halide result from the following allylic rearrangement (see Sect. 2.3.5):

Br + Mg + H–C(=O)–Ph —— 2) $H_3O^{\oplus}$ → [structure: 2-methyl-1-phenylbut-3-en-1-ol, OH]

Hence the structures of the products obtained from these halides as investigated by Knorr and listed in Table 2.4 have been corrected accordingly.

Table 2.4. Results of Barbier-type reactions of allyl bromide or 3-methylallyl bromide with aldehydes and ketones as patented by Knorr [42]. The branch-chained structures of products, obtained with 3-methylallyl bromide (with an asterisk (x)), have been corrected (see page 25). (n.r.: no yields reported)

Carbonyl compound	Product	Yield %
		n.r.
		49
		60
		n.r.
		67
		70
		n.r.
		n.r.
		60
		65
		50
		68

Table 2.4 (continued)

Carbonyl compound	Product	Yield %
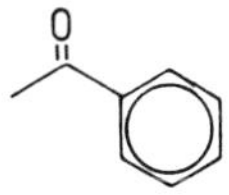	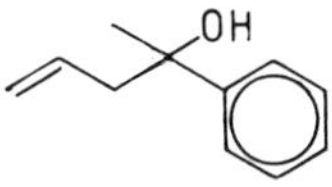	75
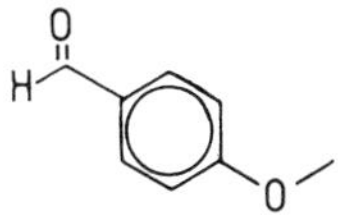	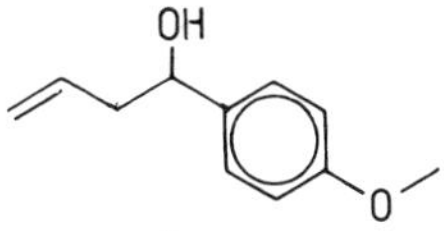	n.r.
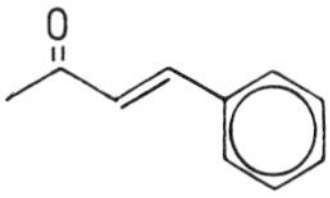	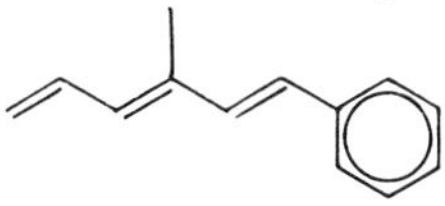	24
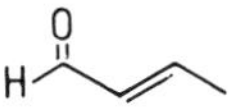 ×	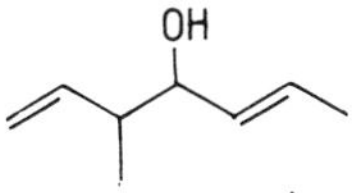	n.r.
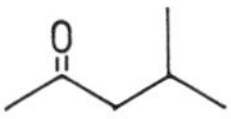 ×	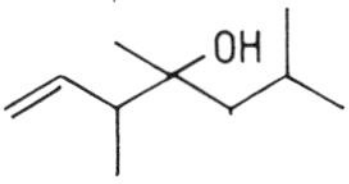	n.r.
 ×	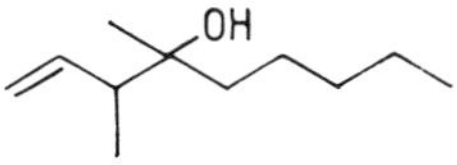	n.r.
× 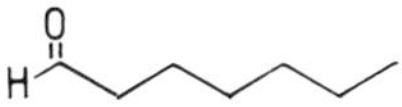	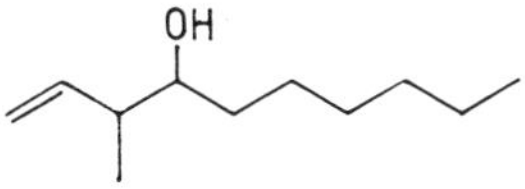	60
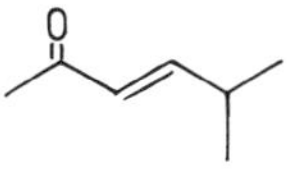 ×	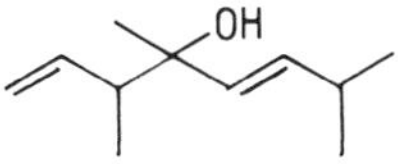	n.r.
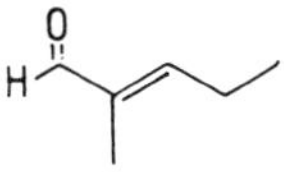 ×	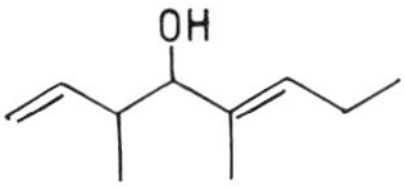	n.r.
× 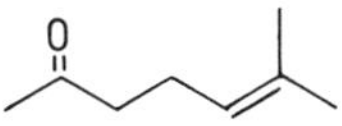	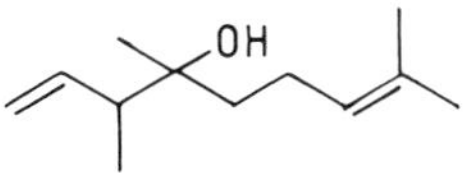	57
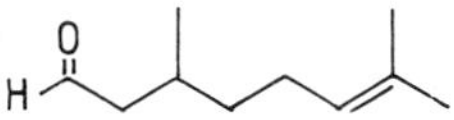 ×	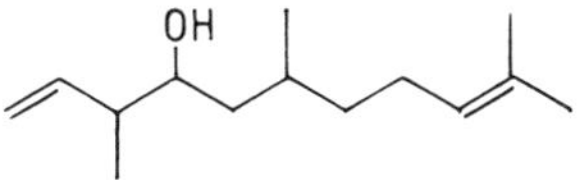	n.r.
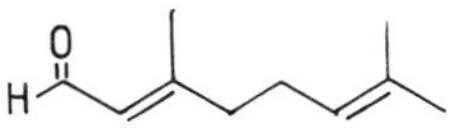 ×	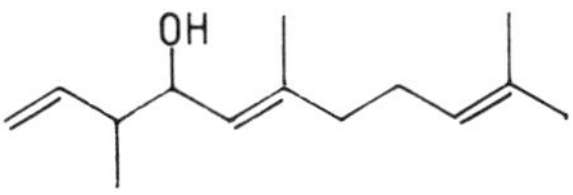	n.r.

Table 2.4 (continued)

Carbonyl compound	Product	Yield %
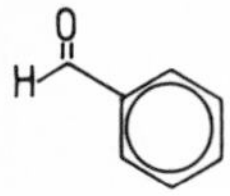 ×	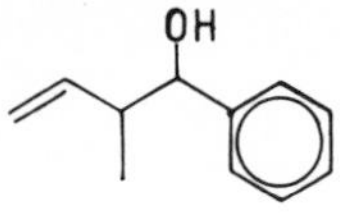	n.r.
× 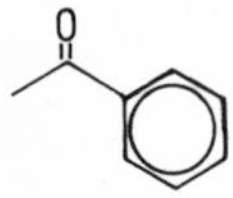	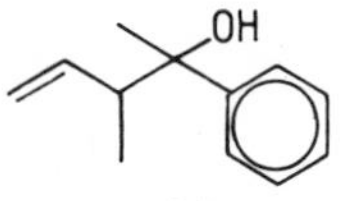	50
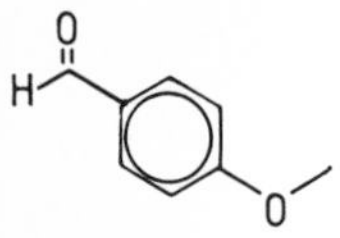 ×	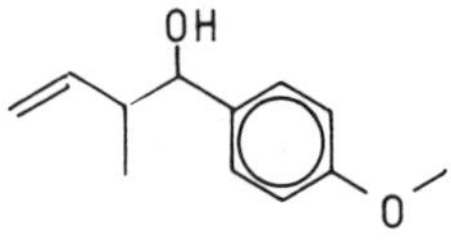	n.r.

Miscellaneous; 1932–1942. The one-step reaction was applied in attempts to synthesize compounds similar to Vitamin A [43].

The reaction of allyl bromide with magnesium in the presence of α-ionone gave the expected carbonyl addition product in fair yield:

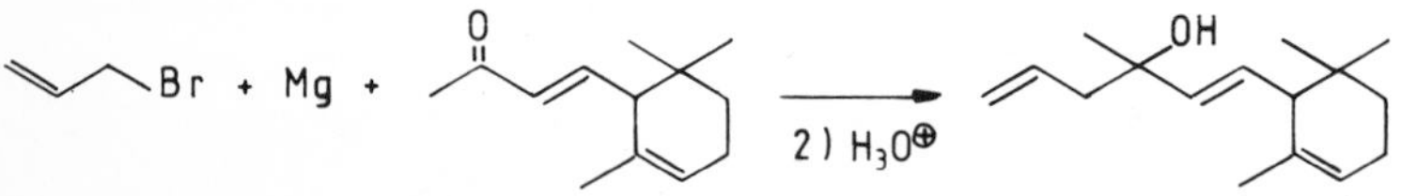

Br + Mg + → 2) $H_3O^{\oplus}$ OH

65%

With the isomeric *β*-ionone, only Michael addition products were obtained:

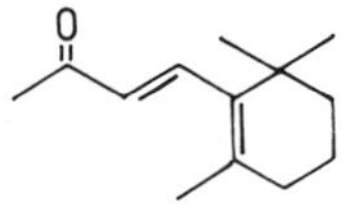

β - ionone

Other researches in the field of Vitamin A like compounds were reported in that same period [44]:

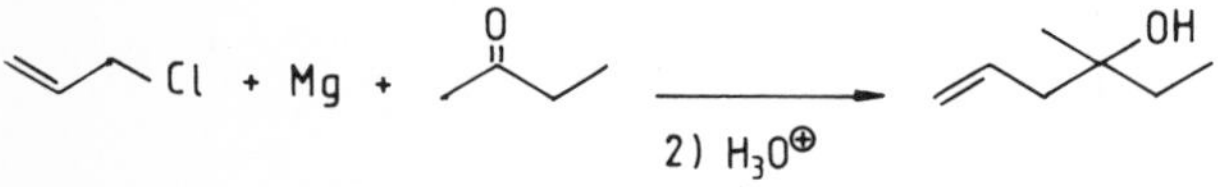

Cl + Mg + → 2) $H_3O^{\oplus}$ OH

52%

and

Br + Mg + → 2) $H_3O^{\oplus}$ OH

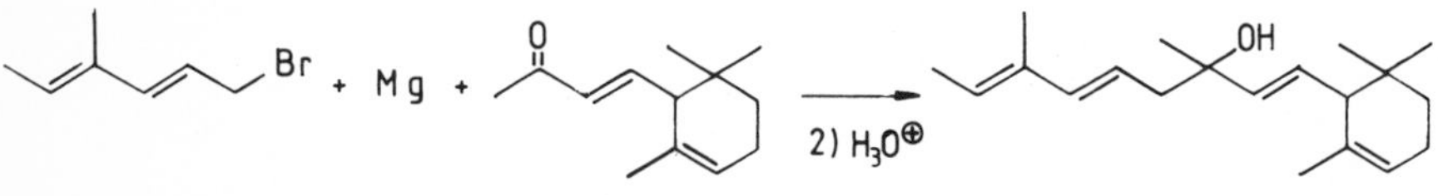

The authors reported no yield of the latter reaction.

The structure of the allylic moiety in the product is very likely to be incorrect (see Sect. 2.3.5).

Surprisingly, in view of the results (81% yield) obtained by Mazurewitsch in 1911 (see Table 2.2), a poor yield was obtained in the following reaction [45] reported at the end of the 1930s:

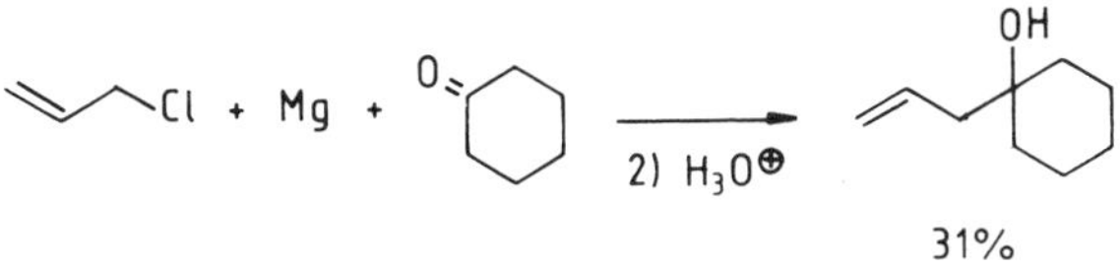

Cl + Mg + O= ⟶ 2) $H_3O^{\oplus}$ OH

31%

No yields were reported in the following Wurtz-type reaction:

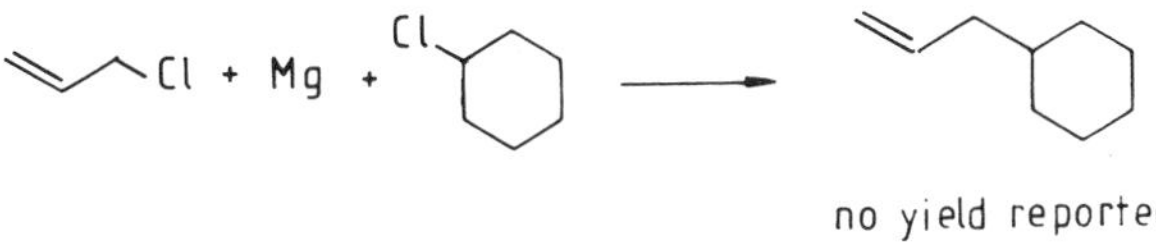

Cl + Mg + Cl ⟶

no yield reported

In the same period quite some criticism came on the application of the Barbier (Jaworsky) reaction [48]:

The preparation of dimethylallylcarbinol from magnesium turnings and an ethereal solution of acetone (1 mole) and allyl bromide (1 mole) as recommended by Jaworsky proved unsatisfactory

(see however again Table 2.2 where yields as high as 75% were obtained for the same reaction as reported by Jaworsky in 1908 and 1909):

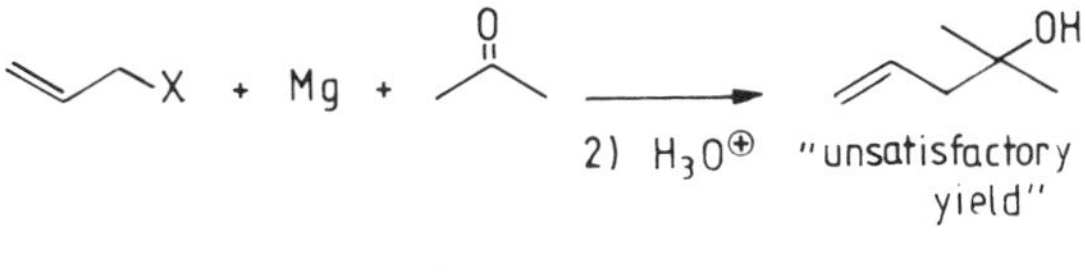

X + Mg + O ⟶ 2) $H_3O^{\oplus}$ OH

"unsatisfactory yield"

On the other hand, the same authors reported, that the two-step procedure

which incorporates Gilman and McGlumphy's precautions (. .) *was used with invariable success*

The cause of such differing opinions on the virtue of the one- and two-step procedures in unclear.

Henze, Allen and Leslie; 1942. The time seemed to have come for the final demise of the Barbier procedure. In 1942, Henze, Allen and Leslie – as real protagonists of the two-step synthetic route – published the results of their investigations in which the Gilman–McGlumphy reagent was used.

A comparison of yields obtained in this study with those obtained in 'batch' procedures, involving allyl halides, a carbonyl compound and zinc or magnesium is unfavourable to the latter (and older) method

were their final words in a ten-page publication in the prestigious Journal of Organic Chemistry [47].

This seemed the end of the Barbier reaction.

For sake of completeness their results are presented in Table 2.5.

In spite of such firm statements, positive results with the one-step procedure with allylic halides were reported only a few years later [48] (although the author was unlikely to have read Henze, Allen and Leslie due to the hostilities at that time):

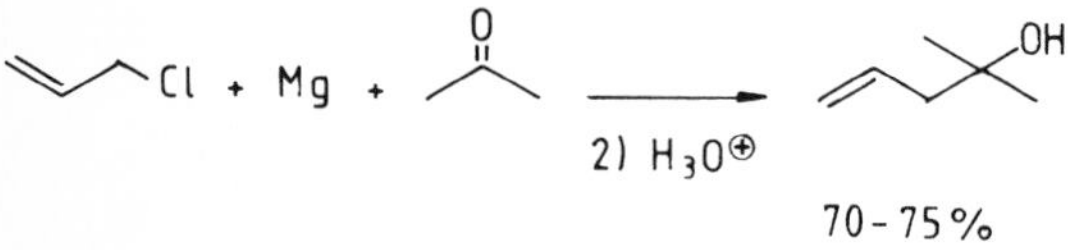

Cl + Mg + O → 2) $H_3O^{\oplus}$ OH

70-75%

Similar positive results with the following reactions were published in that same period [49a]:

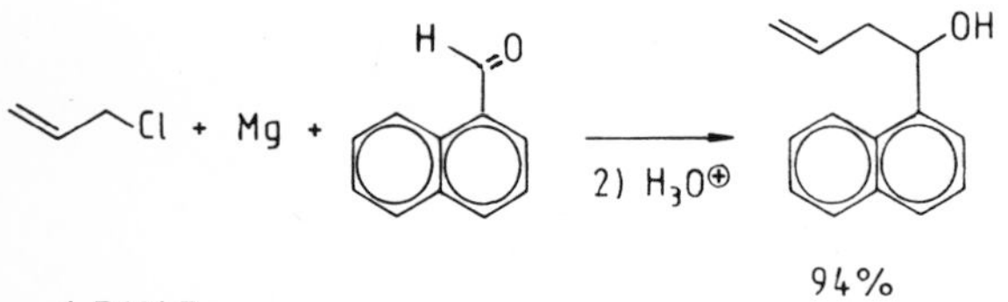

Cl + Mg + H O → 2) $H_3O^{\oplus}$ OH

94%

and [49b]:

Cl + Mg + O → 2) $H_3O^{\oplus}$ OH

Cl + Mg + O → 2) $H_3O^{\oplus}$ OH

Cl + Mg + O → 2) $H_3O^{\oplus}$ OH

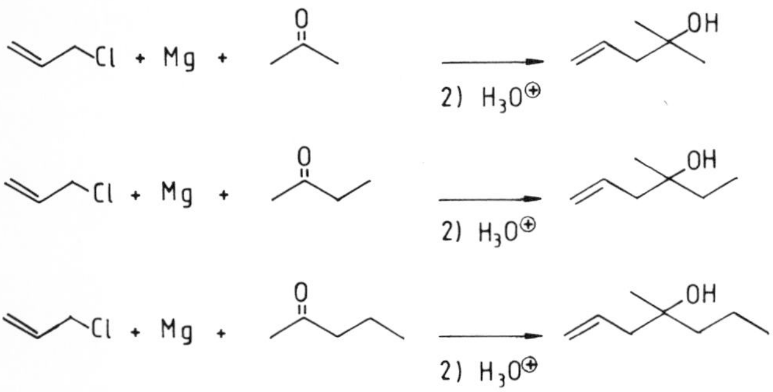

Dreyfuss; 1963. Does the synthetic organic chemist have an intuitive preference for a one-step reaction over a more complicated two-step procedure?

Or is the opposite true – a feeling that things should not be made too complex by the addition of several reagents simultaneously to the reaction mixture; i.e. preference for the two-step method?

Table 2.5. Yields of reactions with allylmagnesium bromide as prepared according to the Gilman–McGlumphy method [47] reported by Henze, Allen and Leslie

Substrate	Product	Yield (%)
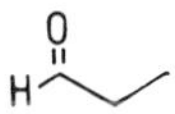	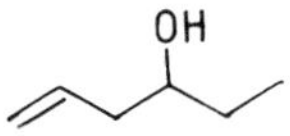	62
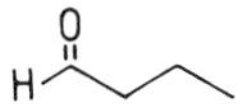	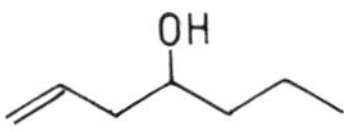	65
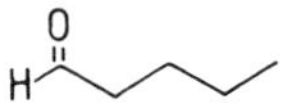	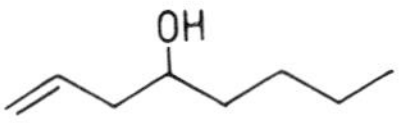	66
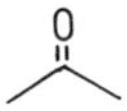		53
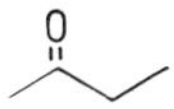		85
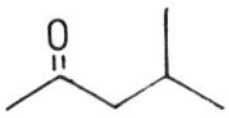	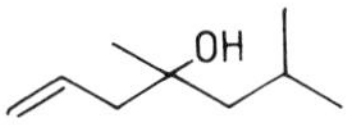	83
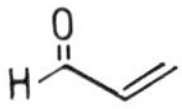	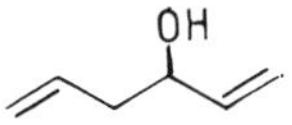	58
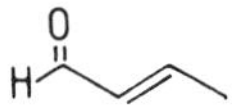	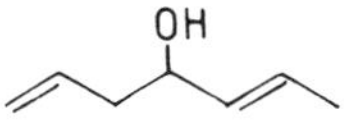	82
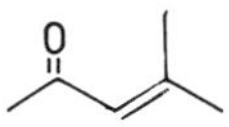	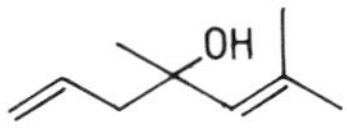	91
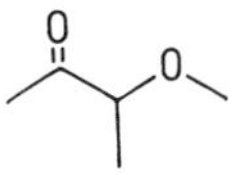	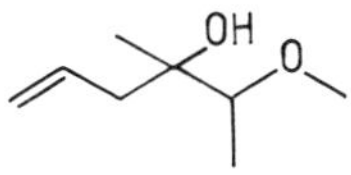	70
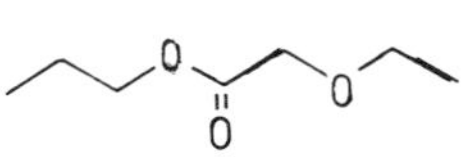	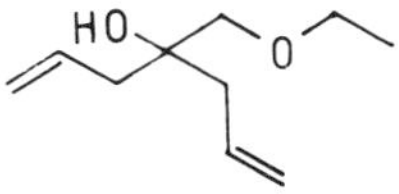	90
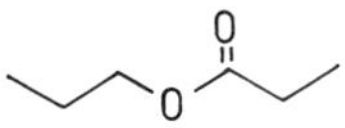	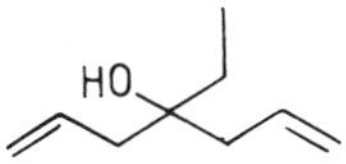	66

Except for the successes of Jaworsky in the beginning of this century and the industrial patent by Knorr some twenty years later no other promising results with the Barbier reaction were published until recently.

It is true that Davies and Kipping expressed their preference for the Barbier-type reaction as early as 1911 [50] (Sect. 1.4) but rapid developments in organomagnesium chemistry seemed to have convinced the majority of chemists that synthesis was best performed in a two-step fashion.

This conviction found further strong support in Gilman's tremendous authority in the field of organometallic chemistry. His method to prepare allylic Grignard reagents eventually led to the commercial availability of allylmagnesium halides in diethyl ether and tetrahydrofuran solutions!

In 1963, approximately fifty years after the introduction of the 'Jaworsky' reaction, Dreyfuss made another attempt to establish the advantages of the Barbier reaction [51].

Following Henze, Allen and Leslie's nomenclature (vide supra), Dreyfuss referred to the Barbier reaction as the Barbier-Grignard procedure and there is little doubt that the conclusion, to be drawn from his interesting publication, is that the one-step procedure for the synthesis of homoallylic alcohols is to be preferred over the two-step Gilman-McGlumphy route.

In Dreyfuss' opinion an essential element for the success of the Barbier reaction is

to start the reaction with a small amount of the allyl halide in ether before beginning the addition of allyl halide and functional addend.

Beside the reactions which he directly studied, Dreyfuss also compared the effectiveness of the two different reaction procedures, i.e. the one-step Barbier reaction and the two-step method via the Gilman-McGlumphy reagent, for the synthesis of 1,5-hexadien-3-ol.

The recommended method for the preparation of this homoallylic alcohol, according to Organic Synthesis [52], is the use of allymagnesium bromide and propenal (acrolein) in diethyl ether, as accepted by the highly respected textbook in 1961:

Br + Mg ⟶ Mg Br

followed by

Mg Br + H(C=O) ⟶ (2) $H_3O^{\oplus}$) OH

By comparing the different reaction conditions of the two methods Dreyfuss convincingly clarified why the one-step procedure is to be preferred. The following scheme presents his results.

Two-Step vs One-Step Preparation of 1, 5-Hexadien-3-ol

	Two-steps [52]	One-step [51]
Moles of magnesium	6.28	2.9
Moles of allyl bromide	2.90	2.5
Moles of acrolein	1.86	2.0
Total volume of diethyl ether	2960 ml	900 ml
Total time of addition of reagents	6 h	2 h
Yield	57–59%	66%

The results of other reactions investigated by Dreyfuss are presented in Table 2.6 and there again show the versatility of what is often named the *Dreyfuss reaction.*

Dreyfuss' article ends with the following remarks:

The Barbier-Grignard method has been shown to be a convenient method for utilizing the allyl Grignard reagent, but it is probably also applicable to Grignard reactions not involving allyl halides. In fact, as long ago as 1911, *Davies and Kippping* [50] *reported the successful application of the Barbier-Grignard method to benzyl and ethyl halides. They also pointed out the need of first starting the reaction with a small amount of halide. Benzyl halide can be thought of as being rather analogous to allyl halide, but the success with ethyl halide suggests the technique may have general application in the Grignard reaction. Indeed it would seem that the work of Davies and Kipping has never received the attention it merits.*

Hartog; 1978. After Dreyfuss' studies on the Barbier reaction with allylic halides (particularly after his final remarks in which he referred to Davies and Kipping's recommendation of the one-step procedure) one would have expected a general acceptance of this simple procedure involving intermediate organometallics.

However, the small number of general studies of the Barbier reaction after 1963 does not seem to confirm that expectation. Several of the more systematic studies that have appeared since then are rather limited in size and come from the patent literature (see e.g. pp. 37 and 59). This suggests that this synthetic procedure was industrially attractive for reasons of its economy (see e.g. Dreyfuss' results as presented in the Scheme above).

Standard laboratory practice indicates that the generally trained synthetic organic chemist is not aware of the advantages of Barbier's method.

The latest contribution to the study of the one-step synthesis with magnesium as the metal appeared at the end of the 1970s [53].

The work illustrates that after some experimenting – in order to find the optimal reaction conditions – Davies and Kipping's conclusions, as well as those of Jaworsky, Knorr and Dreyfuss have to be taken serious.

In this section only the results obtained with allylic halides in one-step reactions will be considered.

Table 2.6. Results of Barbier reactions studied by Dreyfuss [51] using allyl bromide (I) or allyl chloride (II)

Allyl halide	Substrate	Product	Yield (%)
I	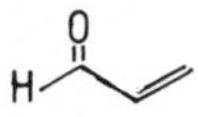	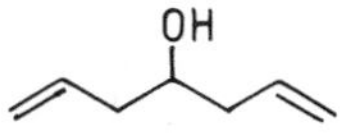	66
I		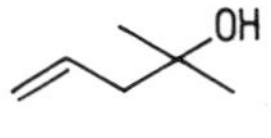	74
II	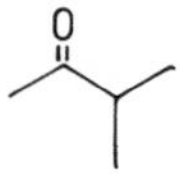	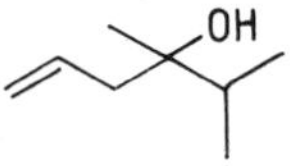	70
I	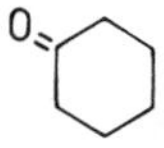	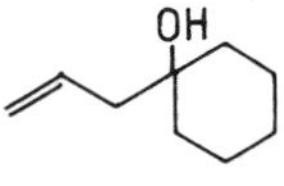	48
I	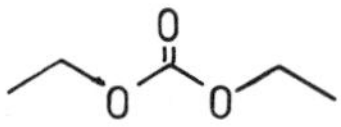	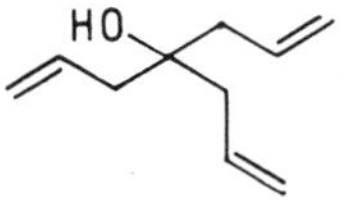	70
I			41
II		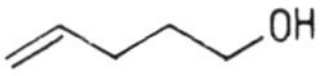	70
I	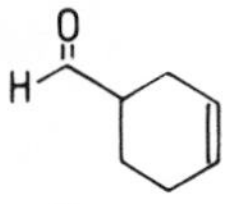	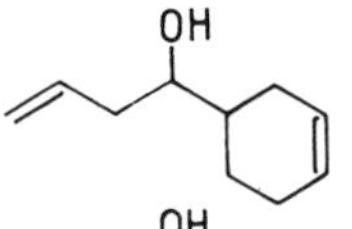	72
II	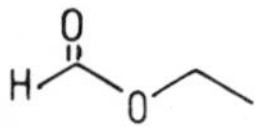	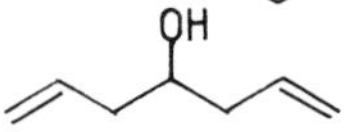	52
I	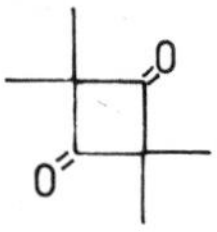	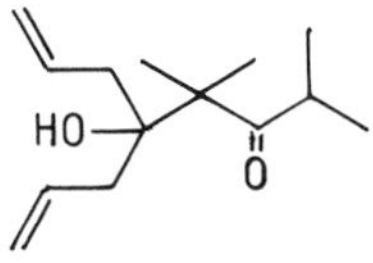	61

The investigations in which other halides were used as well as the studies of the mechanistic aspects of the Barbier reaction will be reviewed elsewhere in this monograph (pp. 60–68 and Sect. 4.4).

The results of Barbier reactions with benzophenone and allylic halides, as presented in Table 2.7, show that the yields are almost quantitative.

Furthermore Table 2.7 shows the yields to be considerably better than those obtained with the preformed Gilman and McGlumphy (Grignard) reagent, in the classical two-step process (see Table 2.3)

Table 2.7. Results of Barbier reactions of allylic halides with benzophenone in different solvents [53]

Allylic halide	Solvent	Product	Yields (%)
 Br	 O	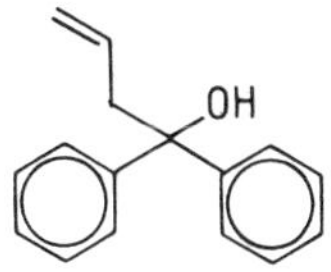 OH	100
"	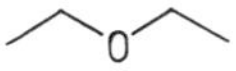 O	"	90
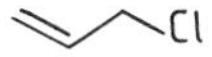 Cl	 O	"	99
"	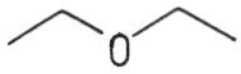 O a)	"	95
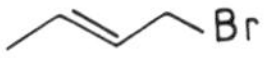 Br	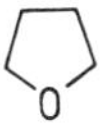 O	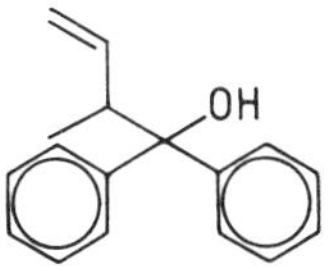 OH	95
"	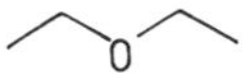 O a)	"	70 b)

[a] a precipitate was formed in the reaction mixture
[b] beside the expected product another product was formed in 20% yield, which contained two 'crotyl' groups, attached to one aromatic ring.

The yields of Barbier reactions of allylic halides with benzaldehyde are also excellent as indicated from the results presented in Table 2.8.

Even when a precipitate is formed – as is the case with diethyl ether as the solvent – yields of 90% can be obtained.

Promising for the industrial application of this type of procedure are the high yields obtained with tetrahydrofuran/benzene mixtures. Even if one molar equivalent of the ethereal solvent was used in benzene, 97% yield of the expected 1-phenyl-3-buten-1-ol could be obtained.

This demonstrates the great advantage of this reaction when large amounts of this highly flammable and relatively expensive solvent have to be reduced.

The next two tables, Table 2.9 and Table 2.10, show the results of Barbier reactions of allylic halides with propanal.

The yields of reactions with allyl chloride and/or bromide are quite good except for those cases where precipitates are formed.

Table 2.8. Yields of 1-phenyl-3-buten-1-ol, obtained in the Barbier reaction of allyl bromide with benzaldehyde in different solvents and solvent mixtures [53]

$CH_2=CHCH_2Br + Mg + H(C=O)C_6H_5 \xrightarrow[2)\ H_3O^{\oplus}]{solvent} CH_2=CHCH_2CH(OH)C_6H_5$

Solvent	Yield (%)
diethyl ether [a]	90
tetrahydrofuran	99
benzene + 2 mol. equiv. tetrahydrofuran	99
benzene + 1 mol. equiv. tetrahydrofuran	97
benzene + 0.2 mol. equiv. tetrahydrofuran	40
benzene	10

[a] a precipitate was formed

From the results, presented in Table 2.10, it can be concluded that Barbier reactions with crotyl halides are much less promising when propanal is used as the second reagent.

It was found that the main sideproduct in these reactions is the tertiary alcohol which contains two crotyl entities. This most probably originates from an addition reaction of the intermediate crotyl Grignard reagent to a ketone, formed in a preceding radical type reaction in which a hydrogen is abstracted from the aldehyde.

This phenomenon will be further discussed in Chap. Three.

$$R\text{-}X + Mg + H(C=O)CH_2CH_3 \longrightarrow R(C=O)CH_2CH_3 + ?$$

Under special reaction conditions (crotyl chloride, tetrahydrofuran as the solvent) however the yield of the desired product can nevertheless be as high as 90%.

Table 2.9. Yields of 5-hexen-3-ol, obtained in the Barbier reaction of allyl halides, magnesium and propanal in diethyl ether and tetrahydrofuran respectively [53]

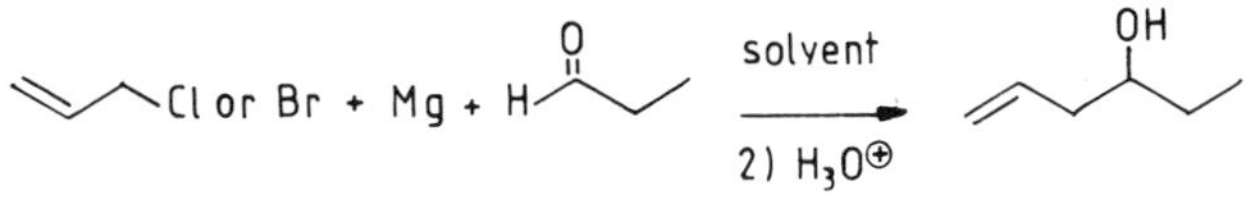

Cl or Br + Mg + H(C=O)CH$_2$CH$_3$ → (solvent; 2) $H_3O^{\oplus}$) → 5-hexen-3-ol (OH)

Allyl halide	Solvent	Yield (%)
 allyl Cl	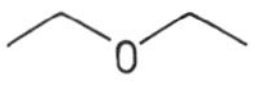 diethyl ether [a]	60
 allyl Br	"	96
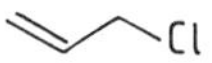 allyl Cl	 tetrahydrofuran	97
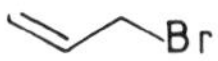 allyl Br	"	98

[a] a precipitate was formed

Finally it was found that Barbier reactions of allylic halides with esters also gave excellent results.

Table 2.11 presents results of reactions of allyl chloride or allyl bromide with ethyl ethanoate in diethyl ether, in tetrahydrofuran or in a mixture of tetrahydrofuran with benzene. All yields were high.

The reactions of crotyl halides with ethyl ethanoate in the same solvents (mixtures) also gave excellent yields, as can be concluded from the results presented in Table 2.12.

Surprisingly relatively large amounts (20% to 25%) of products containing an unbranched crotyl moiety were obtained from these reactions.

Also in the reactions of crotyl chloride with magnesium and diethyl carbonate (see Table 2.13) products were formed containing unbranched crotyl entities, albeit in lower yields: 13–16%. High yields of the expected products were observed.

Japanese Patent; 1990. To demonstrate the versatility of the allyl Barbier reaction mention is made here of a very recent Japanese patent (published in 1990).

A method is claimed [54] for the preparation of 1,1-diphenyl-butadiene derivatives for intermediates for near infrared-absorbing leuco dyes in heat-sensitive recording; the following example is given:

Table 2.10 Yields of 4-methyl-5-hexen-3-ol, obtained from the Barbier reaction of crotyl halides, magnesium and propanal in diethyl ether and tetrahydrofuran [53]

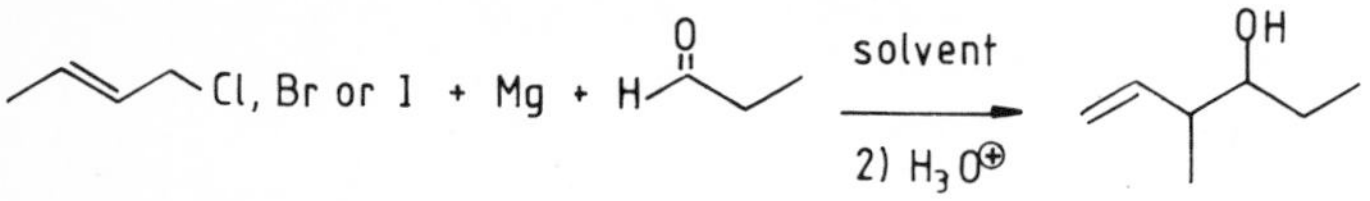

Cl, Br or I + Mg + H(C=O) —solvent, 2) $H_3O^{\oplus}$→ OH

Crotyl halide	Solvent	Yield (%)
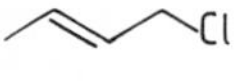 Cl	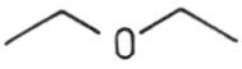 O [a]	50
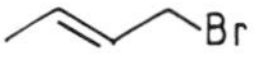 Br	" [a]	50
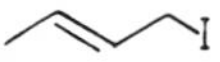 I	"	50
 Cl	 O	90
 Br	"	39
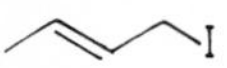 I	" [a]	– [b]

[a] a precipitate was formed
[b] 15% aldehyde dimer was found

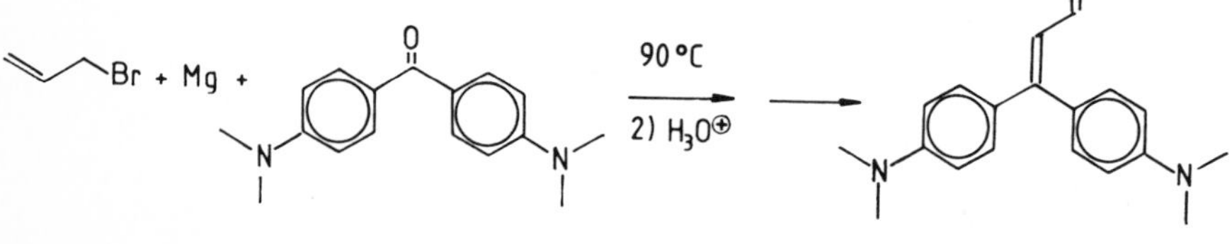

Br + Mg + O, N, N —90 °C, 2) $H_3O^{\oplus}$→ —→ N, N

2.3.5 Allylic Rearrangements

Introduction. In this Section mention must be made of the rearrangements that usually take place when substituted allylic halides are involved in organometallic reactions.

In his German patent [42] on Barbier reactions with allylic halides, Knorr, in 1932, included reactions with crotyl bromide (1-bromo-2-butene)

Table 2.11. Yields of 4-methyl-1,6-heptadien-4-ol, obtained in the Barbier reaction of allyl halides, magnesium and ethyl ethanoate in several different solvents and solvent mixtures [53]

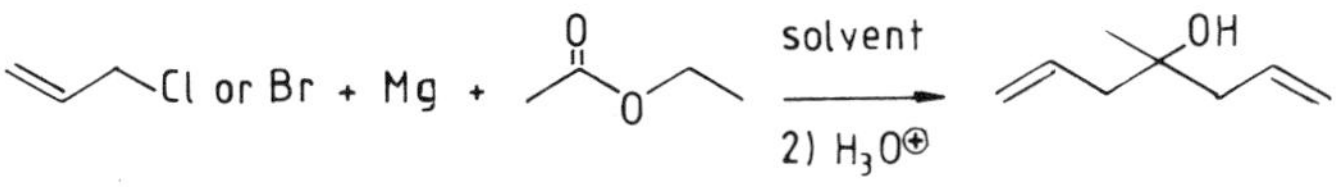

Cl or Br + Mg + ethyl ethanoate —(solvent; 2) $H_3O^{\oplus}$)→ 4-methyl-1,6-heptadien-4-ol (OH)

Allyl halide	Solvent	Yield (%)
 allyl Cl	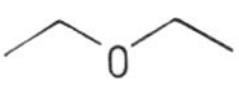 diethyl ether [a]	96
 allyl Cl	 tetrahydrofuran	98
 allyl Cl	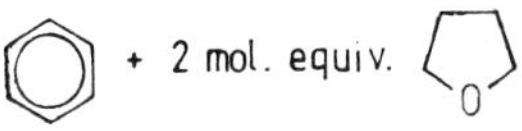 benzene + 2 mol. equiv. tetrahydrofuran	94
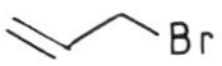 allyl Br	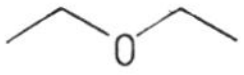 diethyl ether	85 [b]
 allyl Br	 tetrahydrofuran	99
 allyl Br	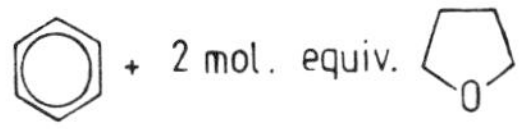 benzene + 2 mol. equiv. tetrahydrofuran	95

[a] a precipitate was formed
[b] 15% ethyl ethanoate was recovered from the reaction mixture

(crotyl bromide: CH3CH=CHCH2–Br)

and assumed that the integrity of the carbon skeleton was preserved in the products.

However, the products of analogous Grignard reagents contain a secondary allyl moiety, suggesting the existence of the following organomagnesium halide:

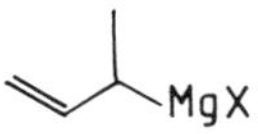

(CH2=CH–CH(CH3)–MgX)

With the development of a method for the preparation of allylic organomagnesium compounds by Gilman and McGlumphy [41] in 1928 (see

Table 2.12. Yields of different tertiary homoallylic alcohols from Barbier reactions of crotyl halides, magnesium and ethyl ethanoate in several different solvents and solvent mixtures [53]

Crotyl halide	Solvent	Yield (%)
crotyl Cl	diethyl ether [a]	a: 59 b: 20
crotyl Cl	THF	a: 63 b: 22
crotyl Cl	benzene + 2 mol. equiv. THF	a: 69 b: 23
crotyl Br	diethyl ether	a: 64 b: 23
crotyl Br	THF	a: 65 b: 23
crotyl Br	benzene + 2 mol. equiv. THF	a: 69 b: 25

[a] a precipitate was formed

Sect. 2.3.3) possibilities were opened up to study the structure and reactivity of substituted allylic organomagnesium compounds.

The following paragraphs will deal with rearrangements observed in allylic organomagnesium compounds.

Since the formation of a secondary Grignard reagent may have been caused by the structural isomerism of the starting primary halide the equilibrium was studied between the two isomers [55]:

1-bromo-2-butene 86% 3-bromo-1-butene 14%

Table 2.13. Yields of different tertiary homoallylic alcohols from Barbier reactions of crotyl chloride, magnesium and diethyl carbonate in various solvents and solvent mixtures [53]

Crotyl chloride + Mg + diethyl carbonate → (1) solvent, 2) $H_3O^{\oplus}$) → product a + product b

Solvent	Yield (%) a	Yield (%) b
Diethyl ether[a]	84	13
THF	76	16
Benzene + 2 mol. equiv. THF	76	16

[a] a precipitate was formed

Equilibrium was reached in 5 days at 20 °C,
in less than 1 hour at 75 °C,
in less than 5 min at 100 °C

The next step in this investigation was the elucidation of the structure of the Grignard reagents, prepared from either one of the two halides.

The mixture of butenes was studied [56], obtained on hydrolysis of such Grignard reagents after most of the solvent had been removed by heating in vacuo at 40–50 °C.

The results indicated that the composition of the gasses (56.4% 1-butene, 26.5% *trans*-2-butene and 17.2% *cis*-2-butene) was independent of the starting bromide.

More than a decade later it was found that the composition of the mixture of butenes depended – among other parameters – on the type of proton donor used [57].

94% 1-butene was formed from 'crotylmagnesium bromide' when treated with 2-phenylethyne

92% was formed when treated with chloroethanoic acid;
80% was formed when treated with ethanoic acid, and
41% was formed when treated with anhydrous ammonium iodide.

It became clear that more sophisticated techniques were required to obtain conclusive information regarding the composition of allylic Grignard compounds in solution.

Ultraviolet spectroscopy [58] did not indicate the existence of a secondary organomagnesium compound in the ethereal solution of the reaction product of 1-bromo-3-phenyl-2-propene with magnesium:

Br + Mg $\xrightarrow{Et_2O}$ MgBr and no MgBr

The same conclusion was reached after infrared spectra of magnesium, zinc and aluminium derivatives of crotyl bromide were studied [59].

On the basis of extensive NMR studies [60] it was concluded that the crotyl Grignard reagent had the primary structure but the possibility of a very rapid equilibration with a small proportion of the secondary reagent could not be excluded.

Young and Roberts, 1946. In a study of reactions of crotyl magnesium bromide, Young and Roberts [61] expressed as their view that the 'abnormal reaction'

should be the expected path for simple allylic Grignard reagents.

This 'abnormal reaction' involved a cyclic mechanism which could account for the formation of branched allylic entities (as well as for *ortho*-substituted phenyl derivatives formed on reaction of benzylmagnesium halides):

Such a cyclic mechanism can explain why the main reaction products of substituted allylic halides contain the secondary allylic system in spite of the primary structure of the Grignard reagent in solution.

The mechanism seems to be generally accepted in organometallic chemistry and is typified by the symbol SE2′ [62–65].

An example of an unexpected reaction of an intermediate allylic organomagnesium compound was reported [66] in a study of the Barbier reaction of 3-methylallyl chloride with propenal.

As would be expected only the branched product was isolated:

Cl + Mg + H (O) —— 2) $H_3O^{\oplus}$ → OH

70%

When, however, this Barbier reaction was executed in much higher concentrations (2–4 molar) the yield of the main product, (4-methyl-1, 5-hexadiene-3-ol) decreased to 60% and a new product, 3, 7-dimethyl-1, 4, 8-nonatriene, was formed in 30% yield.

An electrophilic addition to a carbon carbon double bond was proposed for its formation:

ClMg OMgCl → + MgO + $MgCl_2$

30%

It is of interest to note that the preformed butenyl Grignard reagent does not show this addition reaction to the double bond.

2.4 Other Barbier Reactions

2.4.1 Introduction

The preceding Section has clearly shown that the Barbier reaction is the preferred procedure for syntheses involving allylic halides.

Gilman and McGlumphy's claim [41] – in 1928 – that the use of preformed allylmagnesium bromide gives much better results was convincingly disproved by Dreyfuss' classic publication [51] in 1963 as well as by others (e.g. [53]).

It will be demonstrated in the following Sections that the one-step procedure can also be effective in the reactions of magnesium with halides other than allyl.

First some rather unsystematic researches will be presented in this Introduction which make it clear that the idea of performing syntheses via Grignard reagents in a one-step procedure has come up regularly in the past.

For instance, when attempts were unsuccessful [67] to produce 1, 5-pentanedioic acid by carbonation of the – till then unknown – product of the reaction of 1, 3-dibromopropane with magnesium, a one-step procedure was attempted, just as Houben had done a few years earlier to synthesize 3-butenoic acid [22] (see p. 19). In this case, however, no intermediately formed 1, 3-diorganomagnesium compound could be 'trapped' (see [68] for the preparation of this 1, 3-diGrignard compound); the only acidic compound isolated (in very unsatisfactory yield) was the product of carbonation of a 1, 6-dimagnesium compound formed via consecutive Wurtz-coupling and Grignard reaction:

$$Br(CH_2)_3Br + 2\,Mg + 2\,CO_2 \xrightarrow[2)\ H_3O^{\oplus}]{} HO_2C(CH_2)_6CO_2H$$

very unsatisfactory yield

During a study [69] of alkylation and arylation reactions of naphthalene and biphenyl by organomagnesium compounds it was found that yields of the expected products

tended to be higher when the organomagnesium reagents were prepared in situ than when they were pre-formed in ether or decahydronaphthalene

Together with earlier observations [70] on this type of reactions the conclusion was drawn that

these "in situ" phenomena may involve "nascent" organometallic compounds

In situ reactions of alkyl- and arylmagnesium halides with 6, 6-dialkyl-fulvenes gave, on hydrolysis, substituted fulvenes [71]:

$$\text{6,6-dimethylfulvene} + R\text{-}X + Mg \xrightarrow[2)\ H_3O^{\oplus}]{} \text{C}_5\text{H}_5\text{-C(CH}_3)_2\text{-R}$$

Further reports on syntheses with non-allylic halides are as follows:

2.4.2 Intramolecular Barbier Reactions

The Barbier reaction results in ring formation when the halide and the recipient functional group belong to one and the same molecule:

$$X\text{-}\frown\text{-}Y + Mg \xrightarrow[2)\ H_3O^{\oplus}]{} \text{ring}\text{-}Y'$$

The following paragraphs discuss intramolecular Barbier reactions with different recipient functional groups:

Haloketones in Barbier Reactions. The below introduction in an article on a newly developed cyclization reaction was published in 1970 [72];

Although the intermolecular addition of an organometallic reagent to a carbonyl group is one of the most powerful and useful synthetic methods in organic chemistry, the intramolecular counterpart has never been developed to a useful level . . .

Among the first chemists who recognized this powerful application of the Barbier reaction were Zelinsky and Moser [73], in 1902. These authors expected that in the reaction of a haloketone with magnesium the intermediately formed keto-substituted Grignard reagent would undergo an internal carbonyl addition reaction leading to the formation of a cyclic alcohol (on hydrolysis). Interestingly, they also demonstrated that zinc did not accomplish the same reaction.

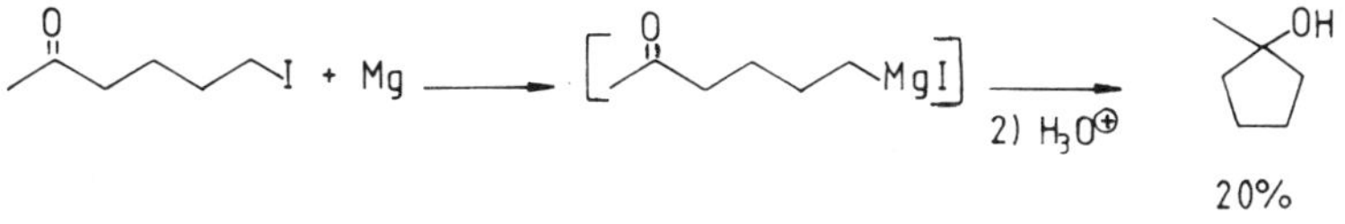

More than sixty years later this Barbier-type cyclization reaction was repeated [74] with the corresponding bromide; methylcyclopentanol was isolated in 65% yield.

The same authors were also successful in the analogous reaction of 4-keto-1-bromoalkanes (see also [75]) resulting in the formation of cyclobutanols. The yields of the reaction (30–60%) were found to be dependent on the length of the alkyl chain

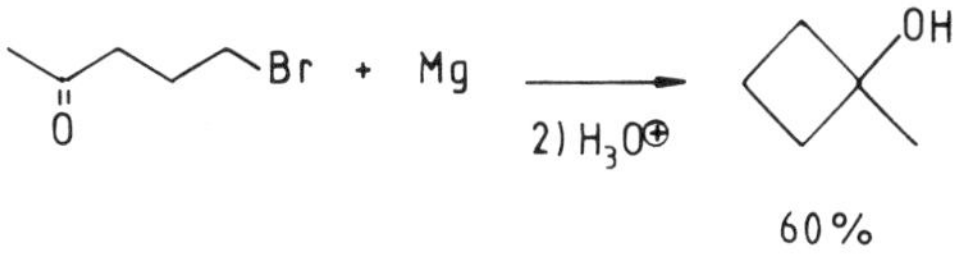

It may be of interest here to report that attempts to prepare 'patchouli-alcohol' via an intramolecular Barbier reaction of a haloketone with magnesium were unsuccessful [76], [77] but that executing this reaction in a sealed tube with sodium in THF at 100 °C gave the required product (see Sect 3.2 for Na-Barbier reactions):

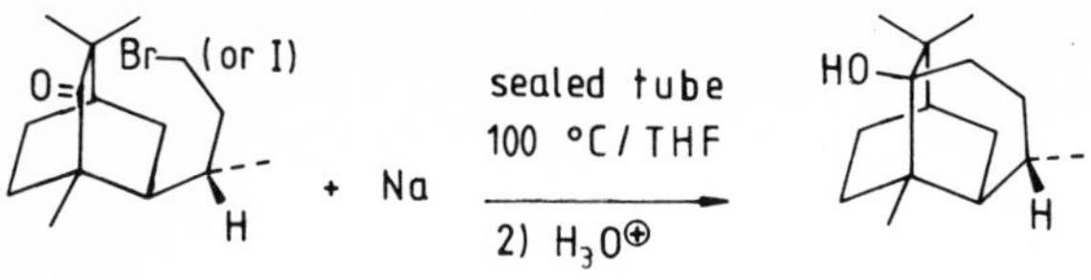

Br (or I) + Na —sealed tube 100 °C / THF, 2) $H_3O^{\oplus}$→ HO

Bicyclo[x.3.0]alkan-1-ols were obtained in moderate to good yields from the same type of reaction when 2-(3-iodopropyl)cycloalkanones were reacted with magnesium in tetrahydrofuran [78]:

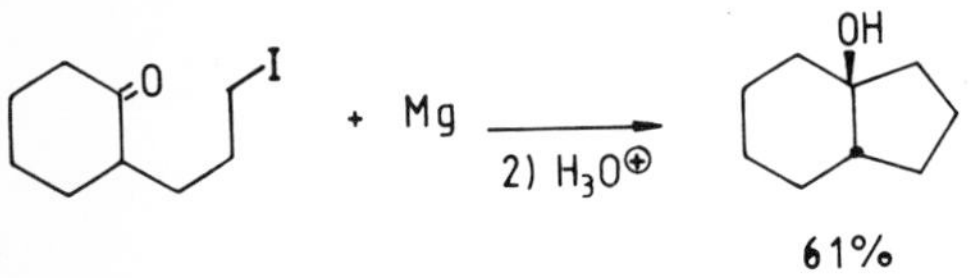

+ Mg —2) $H_3O^{\oplus}$→ OH

61%

Surprisingly, attempts to use the readily available bromide as the starting material were unsuccessful. Also unsuccessful were attempts to extend this reaction to the formation of six- and seven-membered rings.

Studies on the stereochemistry of these intramolecular Barbier reactions in which mono- and tetra-substituted cyclohexanones and several other cycloalkanones were involved revealed a preference for the *cis* ring closure. Some preliminary conclusions were drawn regarding the mechanism of the Barbier reaction (see also Sect. 4.4).

Haloacetals in Barbier Reactions. An interesting variation of the intramolecular Barbier reaction as discussed in the previous paragraph was the use of haloacetals.

An acetal-functionalized Grignard reagent had been used earlier for the synthesis of a keto-carboxylic acid [79]:

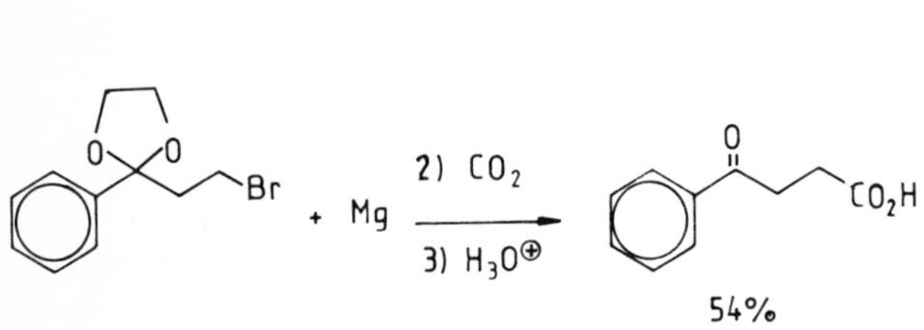

Br + Mg —2) CO_2, 3) $H_3O^{\oplus}$→ CO_2H

54%

Later it was found [80] that, under certain conditions, an intramolecular attack by the Grignard reagent on the acetal function could take place leading to the formation of substituted cycloalkanes:

THF
2) $H_3O^{\oplus}$
63%

With increasing chain-length of the haloacetal, and with THF as the solvent, the exclusive reaction product is the expected Grignard reagent.

However with diethyl ether as the solvent, the ring-scission reaction is much faster and both reaction products may be observed:

Solvent:
THF . . . . sole product

Et_2O . . . . a mixture of these two products

A wide range of reactions of this type with varying chain-length and solvent basicity were investigated.

α-ω-Dihaloalkanes in Barbier-Wurtz Reactions. Barbier-Wurtz cyclization reactions with dihaloalkanes have been performed under many various reaction conditions. Several metals other than magnesium have also been applied in this useful ring-formation reaction (for a recent review see [81]).

Until quite recently it was generally accepted that cyclopropane was the only product from the reaction of 1,3-dihalopropanes with magnesium. However, as much as 30% of the difunctional Grignard reagent could be formed on careful addition of the dihalide to the metal in ethereal solvent [68]:

Mg / Et_2O
main product
31%

Cyclobutane, cyclopentane and higher homologues are much less easily formed [78].

Halonitriles in Barbier Reactions. Barbier reactions with halonitriles will lead to cyclic ketones after hydrolysis of the primary nitrile addition product [82]:

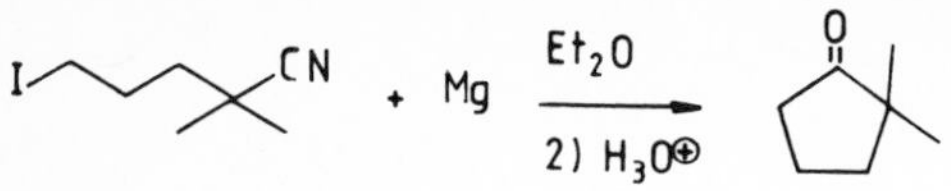

Haloalkylphosphonates in Barbier Reactions. Di-*n*-butyl 5-bromopentyl-1-phosphonate reacted with magnesium (in the presence of magnesium bromide) to form the cyclized product [83]:

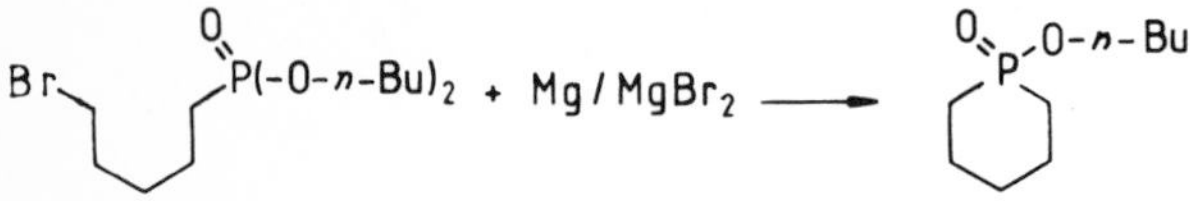

Even more striking is the reaction of a bifunctional halide in a similar process [84]:

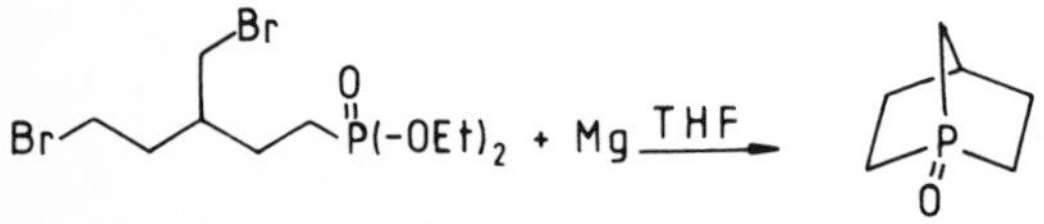

An intramolecular Barbier reaction which might shed more light on the mechanism of this type of reactions is the following one in which aromatic halophosphinates are involved [85]:

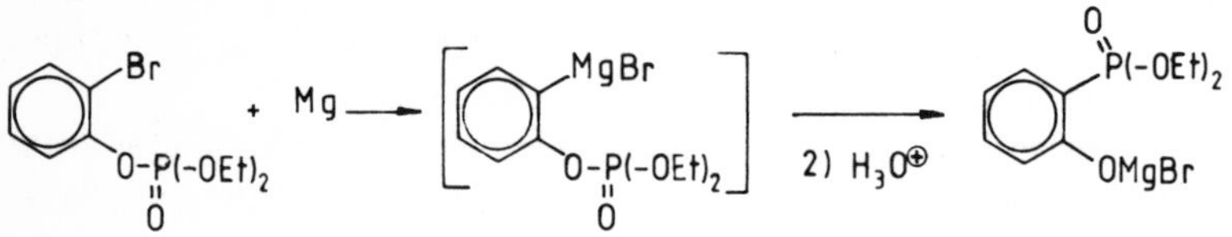

It was proved that this reaction did indeed occur exclusively intramolecularly. Addition, for example, of dimethyl phenylphosphate to the reaction mixture did not result in the formation of the dimethyl ester of phenylphosphonic acid:

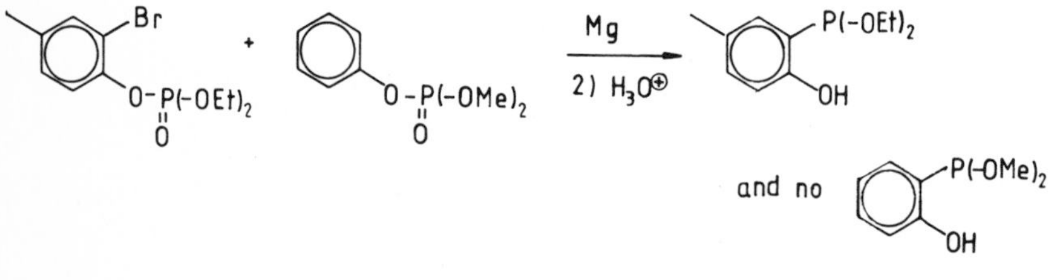

2.4.3 Reactions with Cyclic Acetals

Acetals can be used as the solvent for organometallic compounds but in certain cases a Barbier-type reaction may take place (see Sect. 2.4.2 for intramolecular Barbier reactions with haloacetals).

Tischtschenko, in 1887, first applied acetals in organometallic chemistry in an attempt to develop a general synthetic method for the preparation of primary alcohols [86].

An example is the reaction of diethylzinc with trioxane:

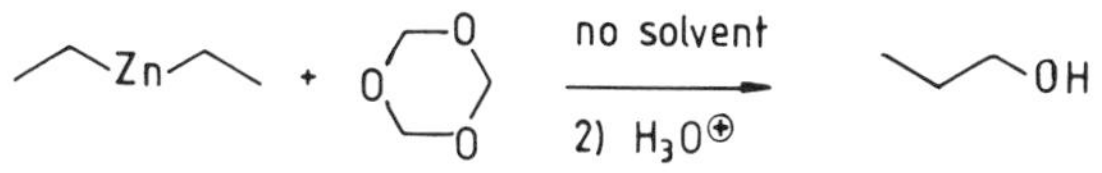

Several years later Wagner and Ginzberg reported the use of 'oxymethylene' in a 'Wagner-Saytzeff'-type reaction with zinc and allyl iodide [87]:

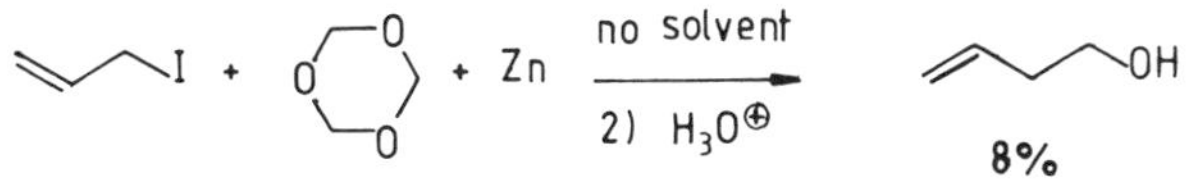

Soon after the discovery of the Grignard reagent, Grignard and Tissier [88] synthesized primary alcohols by the reaction of organomagnesium halides with 'trioxymethylene'.

Fournier, a few years later [89], used the same cyclic acetal in a one-step procedure. No yields were reported of the products which were used as intermediates in the synthesis of carboxylic acids.

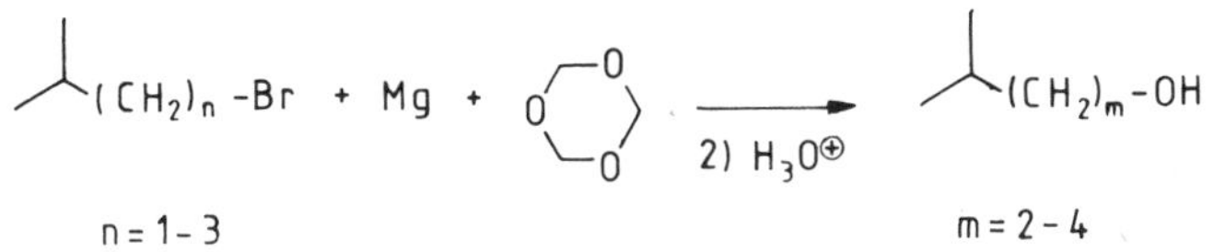

Lespiau's [23] synthesis (in the same period) of 3-buten-1-ol (in 30% yield) by a Barbier reaction with allyl bromide, magnesium and trioxane – one of the earliest reports of Barbier reactions with allylic halides – has been mentioned in Sect. 2.3 (p. 20)

Attempting the same reaction several years later, Pariselle [90] obtained the same product in 25% yield, together with its formal:

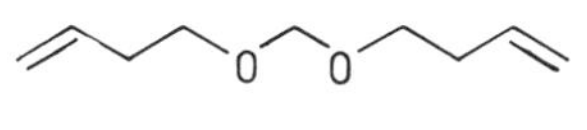

During the early years of Grignard chemistry some of the research was geared towards the use of solvents and/or solvent mixtures, other than diethyl ether in reactions of organic halides with magnesium.

The reaction was studied [91] of iodoethane with magnesium in a hydrocarbon as the solvent in the presence of several different 'catalysts', one of which was paraldehyde, the trimer of ethanal (Oddo and Del Rosso; 1911),

3 EtI + 3 Mg + paraldehyde —(hydrocarbon; 2) $H_3O^{\oplus}$)→ 3 butan-2-ol (OH)

More recent publications [92] in this field report the use of cyclic acetals such as 1,3-dioxolanes as the 'catalysts' for Grignard reagent formation reactions in non-polar solvents such as benzene:

RX + Mg + R'R''-dioxolane —(benzene; 2) $H_3O^{\oplus}$)→ HO–CH₂CH₂–O–C(R)(R')(R'')

Yields were excellent, indicating that oxygen-containing polar organic substrates are good solvents for Grignard reagents in hydrocarbons. As mentioned in the previous Section (p. 46), reactions of haloacetals with magnesium do not always lead to ring-scission [79].

For the mechanistic aspects of these reactions see [92d].

2.4.4 Miscellaneous Barbier Reactions

Davies and Kipping; 1911. Davies and Kipping (see also Sect. 1.4) were the earliest protagonists of the one-step Barbier reaction at a time when the euphoria over the use of Grignard's reagent seemed to have blinded chemists to other synthetic possibilities.

In their early publications on the synthesis of organosilicon derivatives, Kipping and coworkers reacted benzyl chloride with magnesium in the presence of silicon chlorides [93]:

3 $PhCH_2Cl$ + 3 Mg + $SiCl_4$ —(Et_2O)→ $[PhCH_2]_3SiCl$

The success of this one-step procedure inspired Kipping and Davies to extend this type of reaction [50] to prepare

carbinols from aldehydes, ketones and esters, in order to ascertain whether this modification of the usual method could be advantageously applied to other cases

They obtained good results with their one-step procedure (see Table 2.14) and came to the conclusion that

the preliminary preparation of the magnesium compound is unnecessary

Oddo and Del Rosso; *1911.* It was during that same period that research started on the composition and structure of Grignard compounds in solution. In 1906–7, Chelinzeff [94] investigated the role of the solvent by measurement of the heats of solvation for Grignard reagents.

For this purpose solvent-free organomagnesium compounds were prepared by the reaction of organic halides with magnesium in benzene in the presence of 'catalytic' amounts of *N*, *N*-dimethylaminobenzene. These reagents were then placed in a calorimeter, small portions of ethereal solvents were added, and the evolution of heat by solvation determined.

The work of Oddo and Del Rosso [91], (p. 50) was related to Chelinzeff's investigation.

These authors discovered that other polar organic compounds such as aldehydes, ketones, etc. could be used to promote the Grignard reagent formation reaction in apolar solvents thus introducing a new variation on the Barbier reaction in which hydrocarbons were used as the solvent:

I + Mg + O (hydrocarbon; 2) $H_3O^{\oplus}$) → OH

Their results, which they reported to be 'satisfactory', are listed in Table 2.15.

Schlenk; *1931.* A further 20 years went by before more systematic research was done on the application of the Barbier reaction.

Knorr published a patent on the preparation of homoallylic alcohols in 1930 (see Sect. 2.3.4, p. 25 and Table 2.4) and one year later Schlenk studied the preparation of Grignard reagents in hydrocarbon solvents in the presence of polar organic reactants [95].

Iodobutane and iodooctane gave excellent yields of Grignard reagents when reacted with magnesium in benzene but (remarkably so) the yields were poor with iodomethane, -ethane, -propane and -heptane. No comments were given on these results.

Table 2.14. The results of Barbier reactions by Davies and Kipping with bromoethane or benzyl chloride, aldehydes, ketones or esters and magnesium in diethyl ether as the solvent [50]. (n.r. = not reported)

Halide	Substrate	Product	Yield (%)
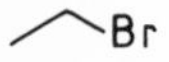 Br	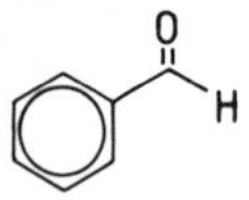 O H	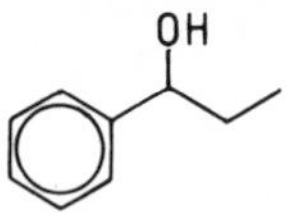 OH	appr. 100
,,	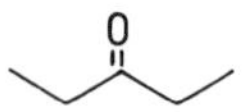 O	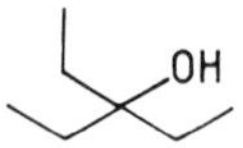 OH	60
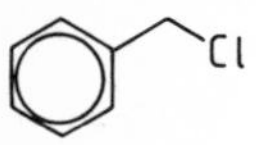 Cl	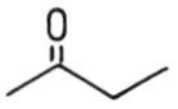 O	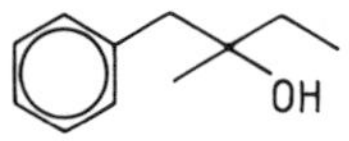 OH	n.r.
,,	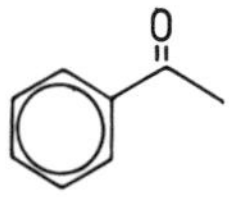 O	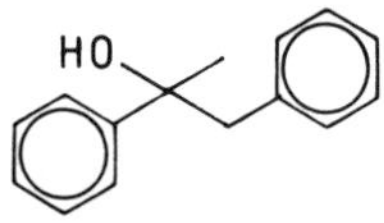 HO	n.r.
,,	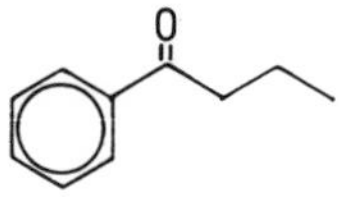 O	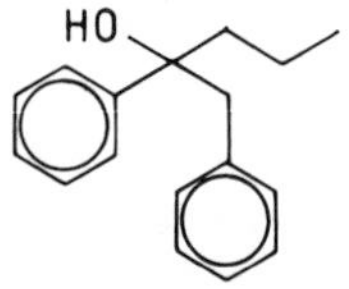 HO	60
,,	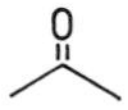 O	no reaction	
,,	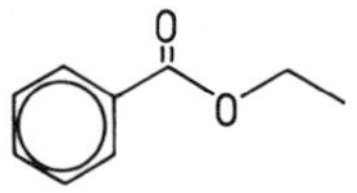 O O	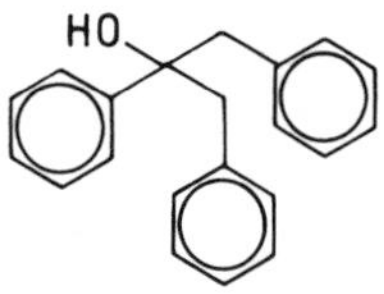 HO	60
,,	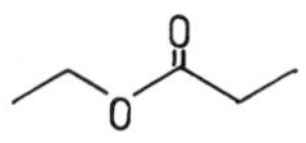 O O	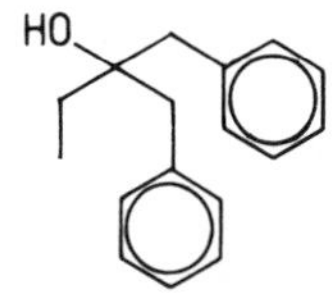 HO	"good"

When this reaction was carried out in the presence of ketones and esters however, the Barbier reaction products were obtained in reasonable yields as can be concluded from the results listed in Table 2.16.

Miginiac-Groizeleau; 1961. Some thirty years ago several new solvents were introduced into organometallic chemistry. The qualities were investigated among others of the weakly basic furan for Grignard reagent formation reactions.

Table 2.15. Barbier reactions between an organic halide, magnesium and an organic substrate with either benzene or petroleum ether as the solvent as studied by Oddo and Del Rosso [91].

Halide	Substrate	Product (s)
I	O	OH
I	,,	OH
,,	O H	OH
,,	O H	OH
I	O H	OH
I	O O	OH
,,	O O O	OH
,,	OH	C_2H_6 + OH
,,	NO_2	H N + N=N

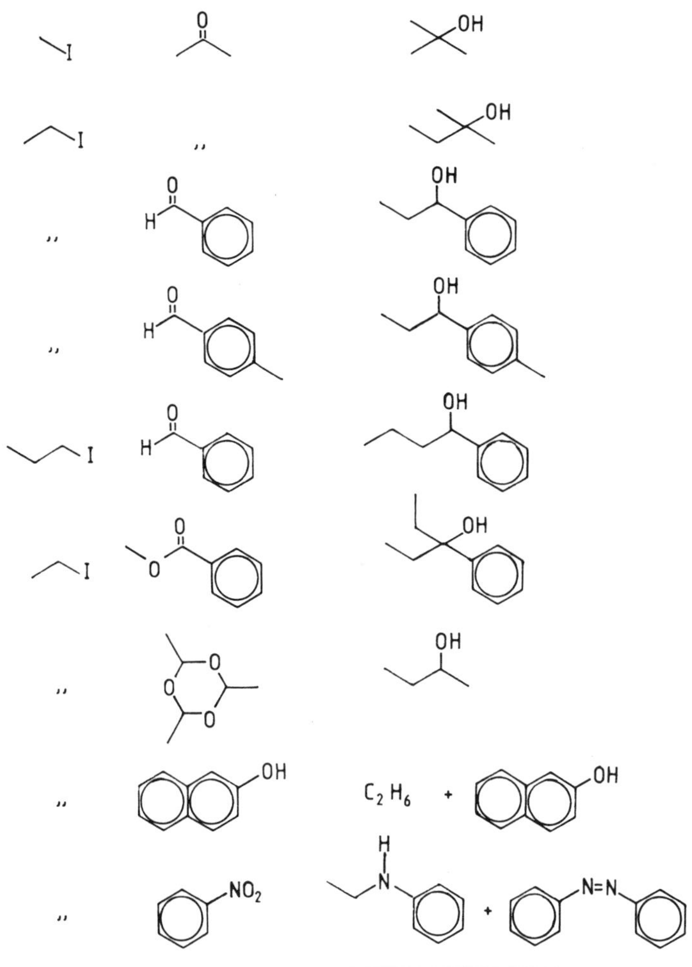

In this solvent the organomagnesium compounds were only slightly soluble and the reaction was very slow [96].

When the reaction was performed in the 'Saytzeff-manner' however, i.e. simultaneous addition of the substrate, results obtained were satisfactory (see Table 2.17) and can be summarized as follows:

a. Organic chlorides react slowly.
b. With organic bromides only small amounts of Wurtz-type coupling reaction products were found.

Table 2.16. Results of Barbier reactions between an organic halide, magnesium and an organic substrate with benzene as the solvent [95] (n.r. = not reported)

Halide	Substrate	Product (s)	Yield (%)
I	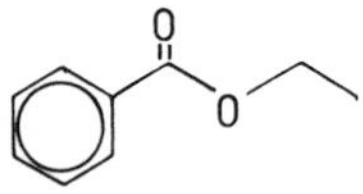 O O	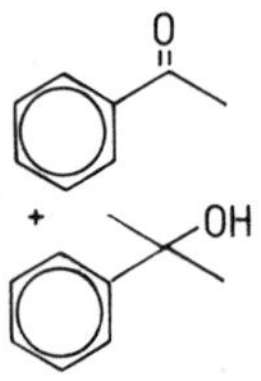 O + OH	17 34
,,	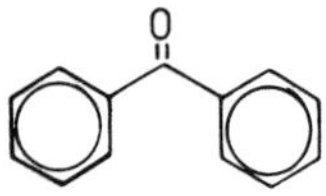 O	-	
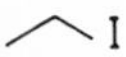 I	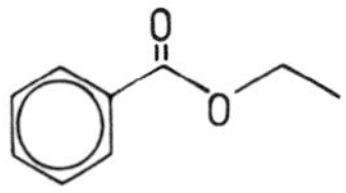 O O	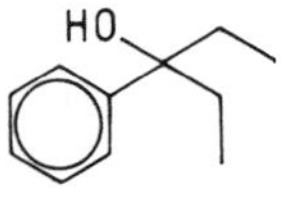 HO	73
,,	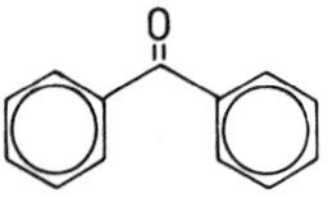 O	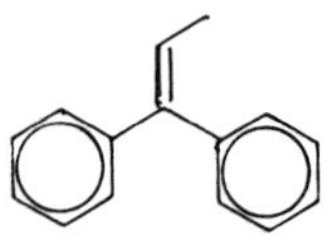	47
,,	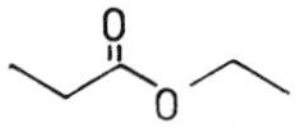 O O	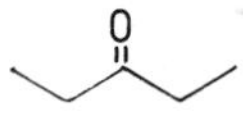 O	58
,,	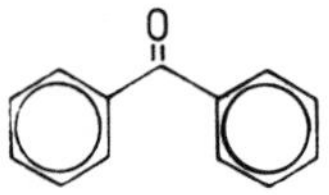 O	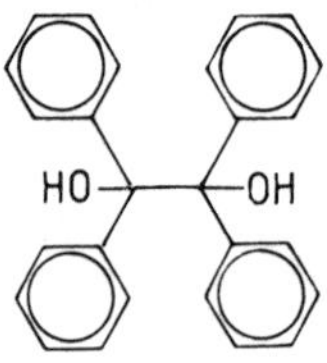 HO OH	n.r.
Br	O O	Br O O	47 30
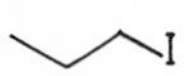 I	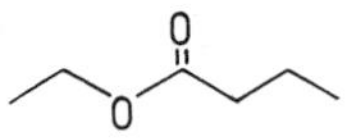 O O	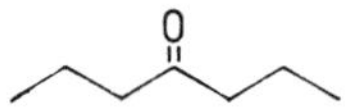 O	52
 Br	,,	-	
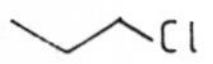 Cl	,,	-	

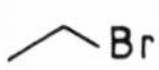

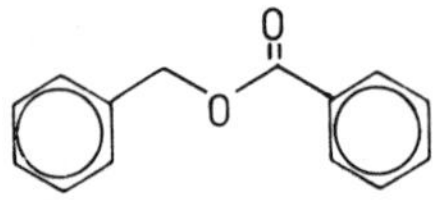

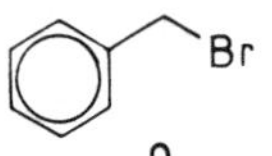

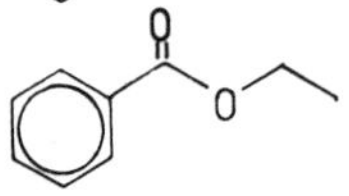

Table 2.16 (continued)

Halide	Substrate	Product (s)	Yield %
I	O, O	+	n.r.
		CH, 3	"mainly"
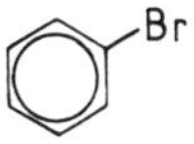 Br	"	-	

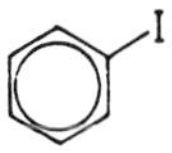

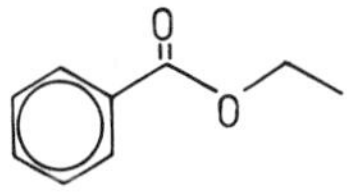

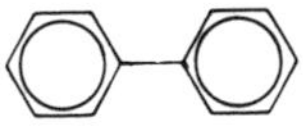

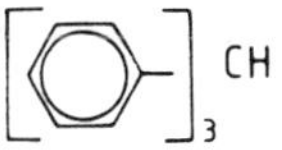

Table 2.17. Results of Barbier reactions with furan as the solvent [96].

Halide	Substrate	Product	Yield (%)
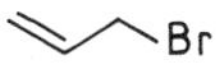 Br	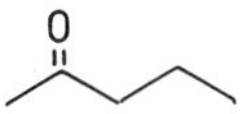 O	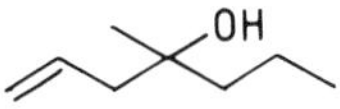 OH	80
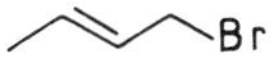 Br	"	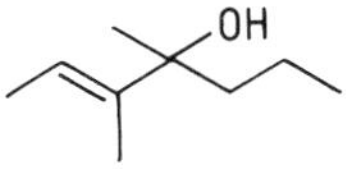 OH	70
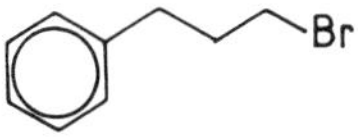 Br	 O	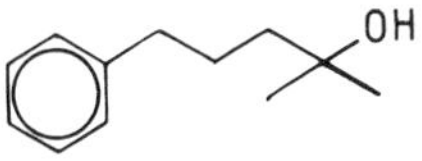 OH	65
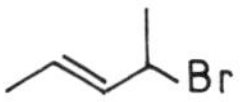 Br	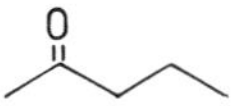 O	no 'normal products'	
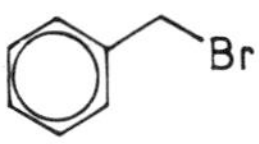 Br	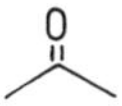 O	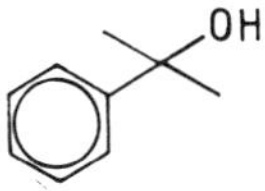 OH	72
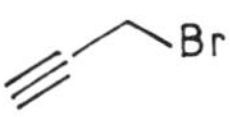 Br	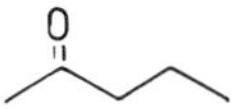 O	 OH	70
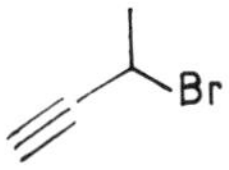 Br	"	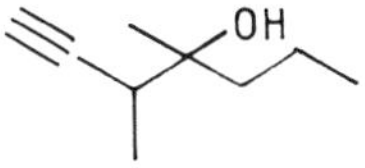 OH	85
 I	"	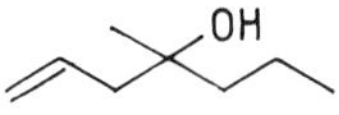 OH	85
		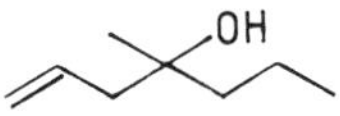 OH	70

Table 2.17 (continued)

Halide	Substrate	Product	Yield (%)
 Cl	,,	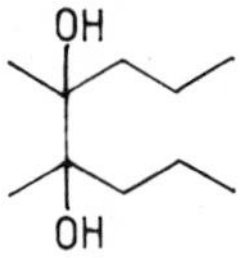 OH OH	25
 Cl	,,	no reaction	
 Br	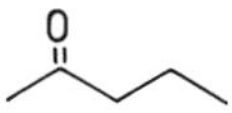 O	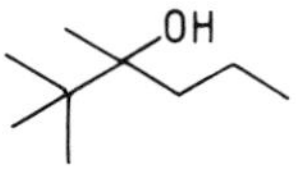 OH	69
,,	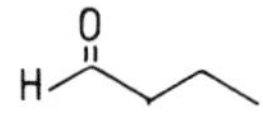 O H	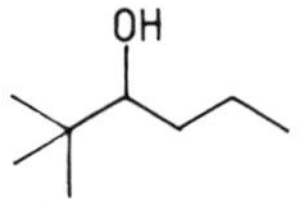 OH	30
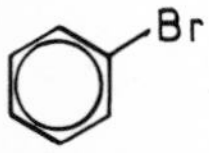 Br	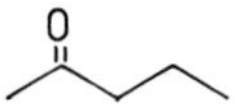 O	no reaction	
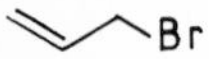 Br	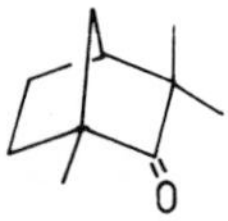 O	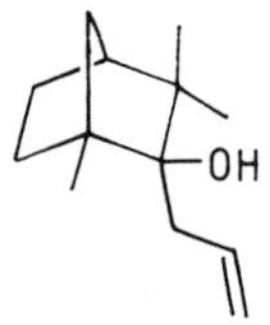 OH	50
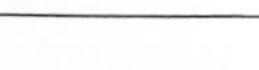 Br	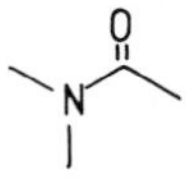 O N	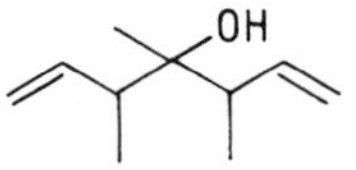 OH	37
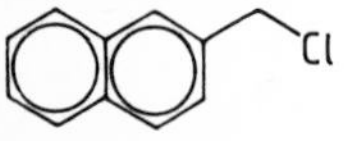 Cl	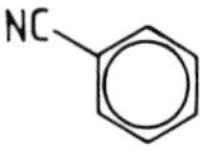 NC	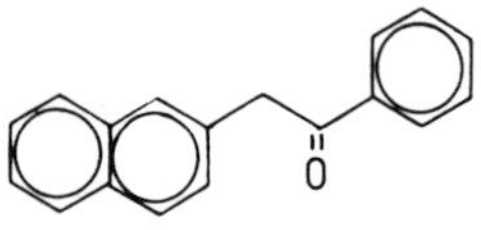 O	"poor"

c. The reactions with carbonyl compounds give good results, in some instances better than with the two-step procedure.
d. There was much less dimerization of the carbonyl compound than usual (pinacol formation).
e. No reduction or enolization reaction products of the carbonyl compounds were found.

In fairness it should be stated that furan cannot be considered an ideal solvent for organometallic compounds. Its low boiling point, 32 °C, makes it very impracticable and its cost is prohibitive for large-scale use.

It may have some interesting features, however, for more theoretical studies of the Barbier reaction.

French Patent; 1964. In a French Patent a method was published for the synthesis of aldehydes on an industrial scale by applying the one-step procedure [97]; objections against the two-step procedure were the following:

a. the preparation and handling of the organomagnesium compounds is rather difficult;
b. the yields are small to moderate;
c. during the second step, i.e. during the condensation of the organomagnesium compound and the reactant, the high concentration of the former compound causes the formation of sideproducts.

In the proposed one-step reaction an excess of the organomagnesium reagent is avoided and the yields obtained are high (almost theoretical yields were claimed).

The patent covers both alkyl and aryl halides and – as the second reagent – *N*-methylformamide, methanoic acid, its esters or its *ortho* esters. The following example is given:

CF3 Br + Mg + N H O → 2) $H_3O^{\oplus}$ CF3 O H

90%

Bifunctional Halides; 1971. Investigating the preparation of 1, 1-bifunctional Grignard reagents use was made of a Barbier-type of reaction in which a 1, 1-dihalide reacted with magnesium in the presence of an aldehyde or a ketone [98] (diethyl ether or an ether/benzene mixture was used as the solvent):

Br Br + 2 Mg + O → Et_2O 2) $H_3O^{\oplus}$

68%

With more complicated molecules the results of this type of reaction were often very satisfactory as can be concluded from the following list:

dodecanal	yielded	65%	tridecene
benzaldehyde	,,	70%	styrene
p-chlorobenzaldehyde	,,	80%	*p*-chlorostyrene
6-undecanone	,,	30%	1, 1-di-*n*-amylethene
benzophenone	,,	40%	1, 1-diphenylethene
5α-cholestan-3-one	,,	70%	3-methylene-5α-chol-estane
pregn-5-en-3β-ol-20-ene	,,	80%	pregn-5-en-3β-ol-20-methylene
androst-5-en-3β-ol-17-one	,,	45%	androst-5-en-3β-ol-17-methylene

The following remarks regarding the Barbier-type of reactions were made: *The yields indicated refer to the in situ procedure. With the use of preformed methylene magnesium halide the yields are generally better and the reaction is cleaner. This latter procedure is, however, less practical if a large scale reaction is performed*

1.1-Dibromo- and 1, 1,-diiodoethane, 1, 1-dibromo-2, 2-dimethylpropane and benzylidene bromide under several reaction conditions did not show any evidence of the formation of a geminal dimetallic species, although in all cases the starting material disappeared. Trying to capture this species immediately after its formation by an in situ reaction (as this procedure was systematically named) with cholestan-5α-3-one, resulted only in traces of the ethylenic compound being isolated. The bulk of the reaction mixture consisted of three products:

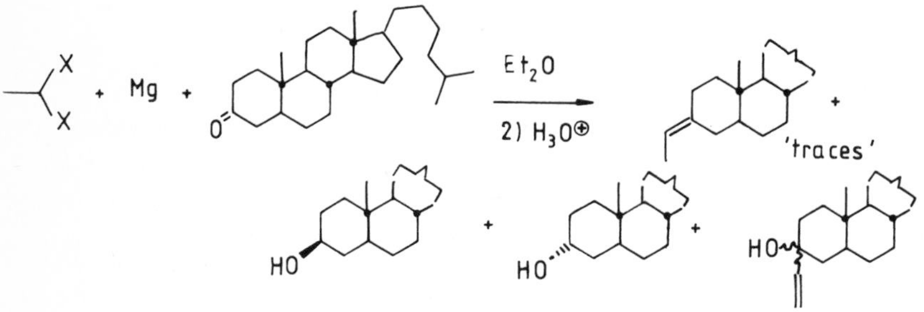

A final reaction studied was as follows:

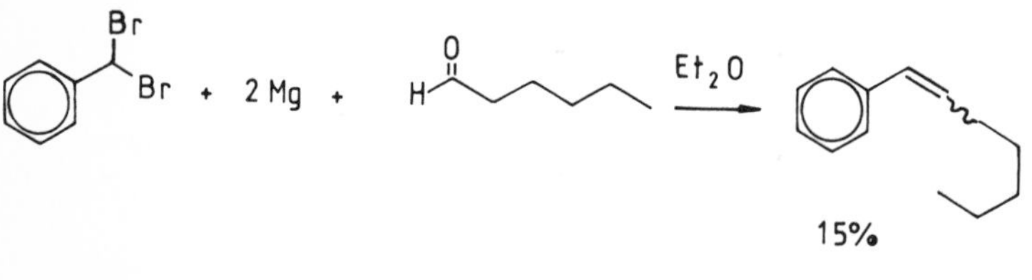

Japanese Studies: 1971 and 1974. Little advantage for the application of the one-step procedure over (what was named) the 'normal Grignard procedure' was found for reactions of butyl halides with heptanal [99].

Although the amount of diethyl ether as the solvent could be reduced by 50% in the one-step procedure and shorter reaction times were required more sideproducts were reported to be found due to Meerwein-Ponndorf-type reduction reactions:

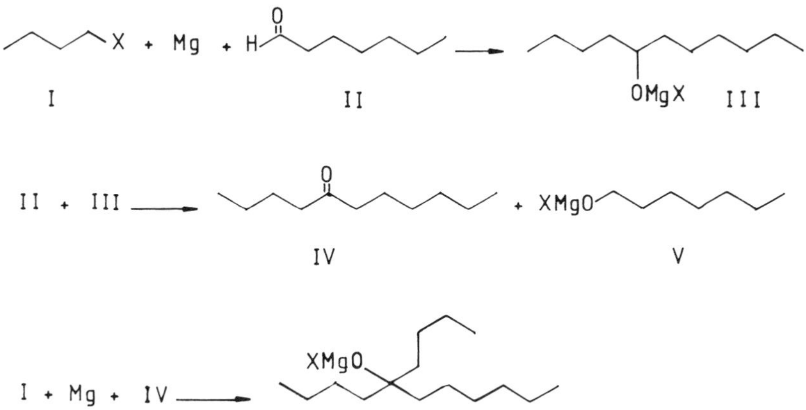

In a Japanese patent, in 1974, the preparation was described of tertiary alcohols by Barbier reactions in cyclic and non-cyclic ether solvents [100].

In general, the patent covered the one-step preparation of alcohols of the following type:

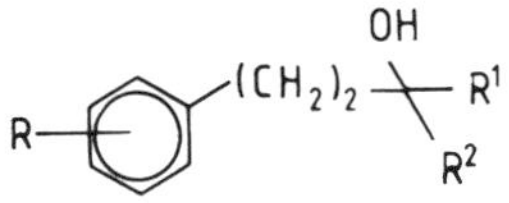

with R^1 and R^2 being C_{1-14} alkyl chains.

Benzyl chloride reacts with propanone to give the expected product in high yield:

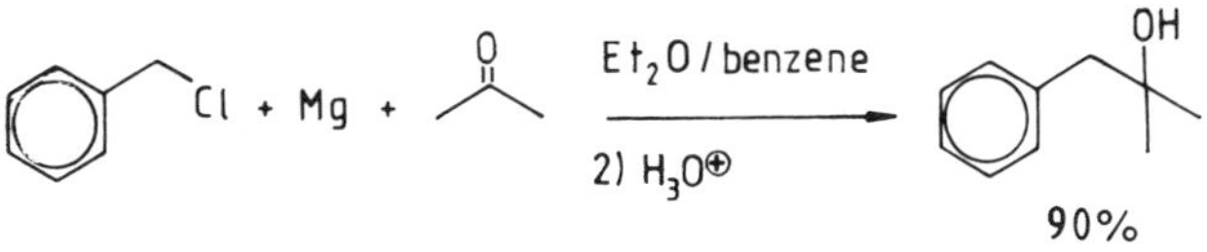

Hartog: *1978*. In Sect. 2.3.4 (Tables 2.7 to 2.13) several results with allylic halides have been mentioned of an in-depth research on the Barbier reaction that took place at the end of the 1970's [53].

Results of other Barbier reactions by the same group are presented in this paragraph and comprise reactions of magnesium and

a) benzophenone	with iodomethane, bromobenzene, bromoethane and -hexane respectively;
b) benzaldehyde	with iodomethane, bromo- and iodoethane as well as with bromobenzene;
c) hexanal	with iodomethane;
d) diethyl carbonate	with bromoethane as well as with bromobenzene;
e) benzyl benzoate	with bromoethane.

The following comments should be made:

a) Benzophenone: Excellent results were obtained in the Barbier reactions with iodomethane and benzophenone in diethyl ether or in a mixture of two molar equivalents of THF in benzene: Table 2.18.

Table 2.18. Yields of products from the Barbier reaction of iodomethane, magnesium and benzophenone in various solvents [53]

I + Mg + (benzophenone) —solvent; 2) $H_3O^{\oplus}$→ A (OH) + B (HO–, –OH) + C,D & E

Solvent	Yield of products (%)		
	A*	B*	other products**
diethyl ether	95	+	C and D
C_6H_6 + 2 mol. equiv. THF	95	+	C and D
THF***	20	+	
C_6H_6***	+	+	E

* + denotes that the products were shown to be present in minor amounts
** C and D: products which show methyl substitution in the aromatic ring
E: tetraphenylethene
See chapter 4.4 for a discussion of the formation of these products
*** a precipitate was formed, which covered the metal and prevented complete reaction

The same is true for reactions with bromobenzene and benzophenone in diethyl ether as well as in THF: Table 2.19. Results of two-step processes compare well with those obtained in one-step reactions.

Much poorer results were obtained in Barbier reactions with bromoethane as the halide and benzophenone.

As can be seen from Table 2.20 the major product was benzhydrol, the product of reduction of benzophenone.

Performing this reaction in the Grignard fashion gave much better results when diethyl ether was used as the solvent, but were again poor in THF.

As was the case with bromoethane, reactions with bromohexane also gave poor results, both in the one-step procedure and in the two-step procedure, reduction of the aromatic ketone becoming the main reaction.

Table 2.21 shows these results.

Table 2.19. Yields of products from Barbier and Grignard reactions of bromobenzene, magnesium and benzophenone in various solvents [53]

C_6H_5Br + Mg + $(C_6H_5)_2C{=}O$ —(solvent; 2) $H_3O^{\oplus}$)→ $(C_6H_5)_3C{-}OH$ (A) + HO–C(C₆H₅)₂–C(C₆H₅)₂–OH (B) + other products: C and D and $C_6H_5{-}C_6H_5$

Solvent/Procedure	Yield of products (%)* A	B	C and D**	$(C_6H_5)_2$
Diethyl ether				
Grignard	90			3
Barbier	90	+	+	6
THF				
Grignard	82			+
Barbier	85	+	+	+
Barbier***	80	+	+	+

* + denotes that the products were shown to be present in minor amounts

** C and D: products which show phenyl substitution in the aromatic ring of benzophenone See chapter 4.4 for a discussion of the formation of these products

*** iodobenzene was used instead of bromobenzene; the solvent was a mixture of benzene and two molar equivalents of THF

Table 2.20. Yields of products from Barbier and Grignard reactions of bromoethane, magnesium and benzophenone in various solvents [53]

Ethyl bromide + Mg + benzophenone → (solvent; 2) $H_3O^{\oplus}$) → A (1,1-diphenyl-1-propanol) + B (benzopinacol) + other products: C and D + F (benzhydrol)

Solvent / Procedure		Yield of products (%)* A	B	C and D**	F
Diethyl ether	Grignard	94			6
	Barbier***	27	19	5	27
THF	Grignard	21	+	+	77
	Barbier	17	10	4	61
	Barbier****	32	+	+	65

* + denotes that the products were shown to be present in minor amounts
** C and D: products which show ethyl substitution in the aromatic ring
See Sect. 4.4 for a discussion of the formation of these products
*** a precipitate was formed; 13% benzophenone was recovered
**** iodoethane was used as the halide; the solvent was a mixture of benzene and two molar equivalents of THF

b) Benzaldehyde: As is the case of Barbier reactions with benzophenone and iodomethane, reactions with benzaldehyde and iodomethane in diethyl ether or in a mixture of two molar equivalents of THF and benzene give excellent yields of the expected products: Table 2.22

Barbier reactions with benzaldehyde and haloethanes were much less successful (as was the case with reactions with benzophenone; Tables 2.20 and 2.21).

Table 2.21. Yields of products from Barbier and Grignard reactions of bromohexane, magnesium and benzophenone in various solvents [53]

$$n\text{-}C_6H_{13}Br + Mg + Ph_2C{=}O \xrightarrow[2)\ H_3O^{\oplus}]{\text{solvent}} \underset{A}{Ph_2C(OH)C_6H_{13}} + \underset{B}{HO{-}CPh_2{-}CPh_2{-}OH} +$$

other products C and D

$$\underset{F}{Ph_2CH{-}OH} + \underset{G}{n\text{-}C_{12}H_{26}}$$

Solvent / Procedure	A	B	C and D**	F	G
Diethyl ether					
Grignard	12			77	11
Barbier***	12	19	5	53	10
THF					
Grignard	3			90	7
Barbier	8	14	1	70	3

Yield of products (%)*

* + denotes that the products were shown to be present in minor amounts
** C and D: products which show ethyl substitution in the aromatic ring
See Sect. 4.4 for a discussion of the formation of these products
*** a precipitate was formed

Table 2.23 shows the results of reactions with bromo- and iodoethane, magnesium and benzaldehyde in diethyl ether and in THF.

In this table products, C, D and E indicate that proton abstraction from benzaldehyde takes place. A discussion of the mechanism of this type of reaction will be given in Sect. 4.4.

Barbier reactions with benzaldehyde and bromobenzene in THF (no reactions were possible in diethyl ether) give excellent yields of the expected products as can be concluded from the results, presented in Table 2.24.

Table 2.22. Yields of products from Barbier reactions of iodomethane with magnesium and benzaldehyde in various solvents [53]

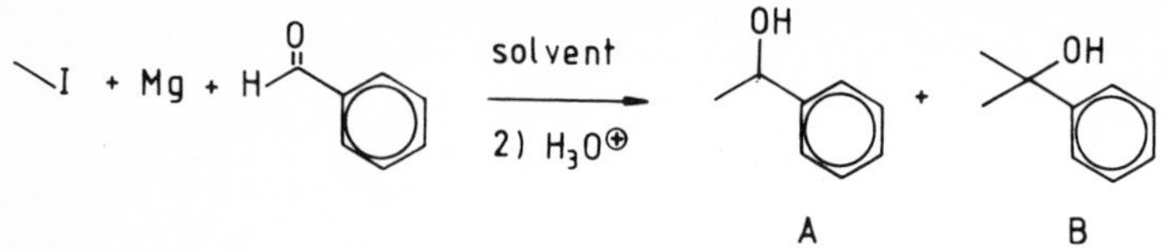

I + Mg + H(O) → solvent, 2) $H_3O^{\oplus}$ → A (OH) + B (OH)

Solvent	Yield of products (%)*	
	A	B
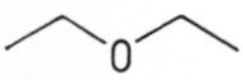 O	95	5
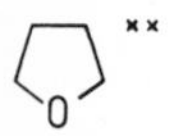 O **	5	+
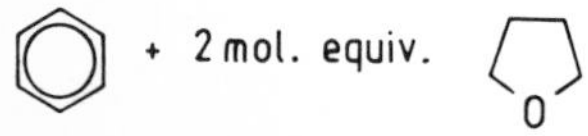 + 2 mol. equiv. O	98	+

* + denotes that the product was shown to be present in minor amounts
** a precipitate was formed which prevented complete reaction

The formation of some unusual products in this reaction will be discussed in Sect. 4.4.

c) Hexanal: Further evidence of the usefulness of the Barbier reaction between iodomethane and aldehydes or ketones is shown in the study of the reaction of hexanal with this halide (Table 2.25). Like in previous cases (see Tables 2.18 and 2.22) a mixture of benzene and two molar equivalents of THF proves to be a more effective medium than THF on its own since in the latter solvent a precipitate is formed.

d) Diethyl Carbonate: Oddo and Del Rosso in 1911, [91] (see p. 50 and Table 2.15) followed by Schlenk in 1931, [95] (see p. 51 and Table 2.16) demonstrated that Barbier reactions with esters lead to good results. This could be confirmed for the one-step reactions with ethyl carbonate and bromoethane or bromobenzene as can be concluded from the results given in Tables 2.26 and 2.27.

e) Benzyl Benzoate: Because of a theoretical interest which will be mentioned in Sect. 4.4, Barbier reactions were studied of bromoethane and benzyl benzoate in various solvents (see also Table 2.16). Yields of the main product, 3-phenyl-3-pentanol, were high as can be seen from the results, presented in Table 2.28.

Table 2.23. Yields of products (in %) from Barbier reactions of bromoethane or iodoethane, magnesium and benzaldehyde in diethyl ether as well as in THF [53]

X + Mg + H–C(=O)–Ph —(solvent; 2) $H_3O^{\oplus}$)→ A + B + C + D + E

Solvent	X	Yield of products (%)*				
		A	B	C	D	E
diethyl ether	Br**	6	3	9	2	+
	I**	15	21	21	3	+
THF	Br	56	24			
	I**	3	15	8	+	5

* + denotes that the products were shown to be present in minor amounts
** a precipitate was formed which prevented complete reaction

Electroassisted Barbier Reactions; 1985 and 1986. More recently a new and efficient method was reported for the *electrocarboxylation* of organic halides, using an undivided electrolytic cell and a sacrificial magnesium electrode [101].

Magnesium was among the metals used for, what the authors – in a following publication – had named an *electroassisted Barbier reaction* [102] (see also Sect. 3.5, p. 129).

The electrochemical reactions were as follows:

$$M - ne \xrightarrow{\text{anode}} M^{n\oplus}$$

$$RX + 2e \xrightarrow{\text{cathode}} R^{\ominus} + X^{\ominus}$$

$$R^{\ominus} + {>}C{=}O \longrightarrow R{-}\overset{|}{\underset{|}{C}}{-}O^{\ominus}$$

Table 2.24. Yields of products from Barbier reactions of bromobenzene, benzaldehyde and magnesium in THF [53]

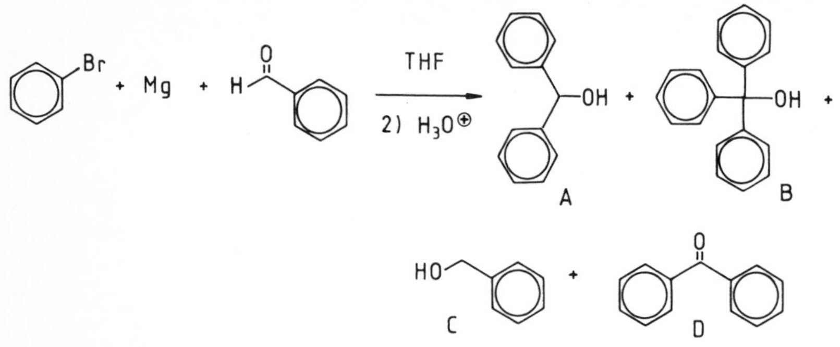

Temp. (°C)	Yield of products (%)*, **			
	A	B	C	D
25	80(37)	2(10)	13	10(17)
45	95	+	–	+
65	96 (10)	+ (20)	– (30)	– (40)

* + denotes that the products were shown to be present in minor amounts
** yields in parentheses were obtained when the reaction was performed carelessly. See Sect. 4.4 for a discussion of the mechanism

Among the halides used were benzyl chloride as in the following example:

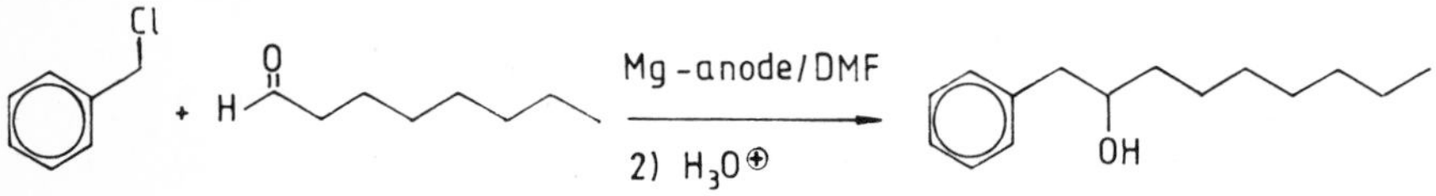

60%

and 1,2,4-trichlorobenzene:

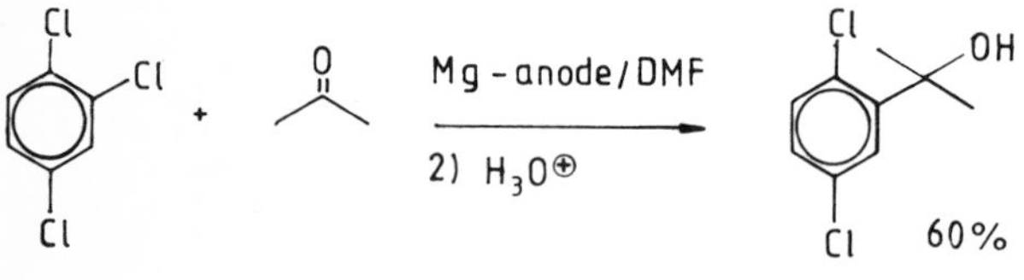

60%

Table 2.25. Yields of products from Barbier reactions of iodomethane with magnesium and hexanal in various solvents [53]

I + Mg + H(C=O)… — 1) solvent, 2) $H_3O^{\oplus}$ → A (OH) + B (OH)

Solvent	Yield of products (%)*	
	A	B
[diethyl ether]	90	5
[THF]**	5	+
[benzene] + 2 mol. equiv. [THF]	95	5

* + denotes that the product was shown to be present in minor amounts
** a precipitate was formed which prevented complete reaction

Table 2.26. Yields of 3-ethyl-3-pentanol from the Barbier reaction of bromoethane, magnesium and diethyl carbonate in various solvents [53]

Br + Mg + [diethyl carbonate] — 1) solvent, 2) $H_3O^{\oplus}$ → HO…

Solvent	[diethyl ether]	[THF]	[benzene] + 2 mol. equiv. [THF]
Yield (%)	100	76	92

Polyhalo Alkanes; 1990. During investigation [103] of the reactions of polyhaloalkanes with magnesium in the presence of a third reactant, using THF as the solvent, a variety of interesting products was obtained. Relevant to this monograph was the observation that benzylidene chloride reacted with magnesium in the presence of benzophenone in a rather predictable way (see [98] p. 57).

Table 2.27. Yields of products from Barbier reactions of bromobenzene with magnesium and diethyl carbonate in various solvents [53]

Solvent	Yield of products (%)		
	A	B	C
diethyl ether	84	5	7
THF	78	10	10
benzene + 2 mol. equiv. THF	86	6	6

Table 2.28. Yields of products from Barbier reactions of bromoethane with magnesium and benzyl benzoate in various solvents [53]

Solvent	Yield of products (%)	
	A	B*
diethyl ether**	84	16
THF	73	27
benzene + 2 mol. equiv. THF	83	17

* Compound B was a 1:1 mixture of 1,2-diphenyl-2-butanol and an isomer, the structure of which was not further investigated
** a precipitate was formed

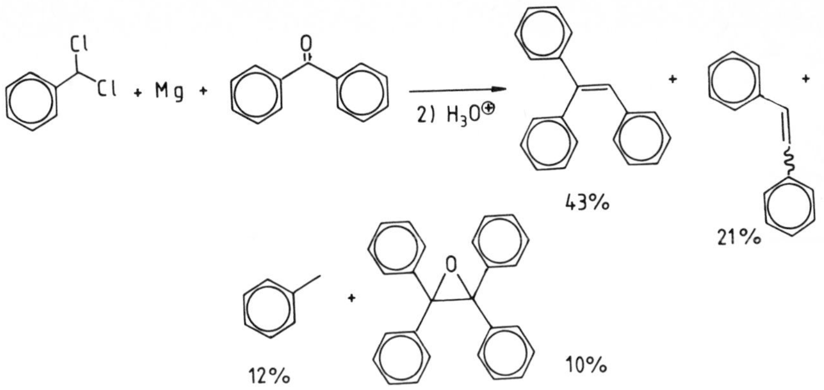

2) $H_3O^{\oplus}$

43%

21%

12%

10%

2.4.5 Barbier Reactions with Halides of Various Elements

One of the first recommendations [50] of the Barbier reaction as the more favourable synthetic route over the two-step procedure via a preformed Grignard reagent was based on researches [93] of one-step coupling reactions of chlorosilanes and benzyl chloride (see p. 50, Sect. 2.4.4)

Numerous coupling reactions with chlorosilanes as well as with other halides of various different elements have, with great success, been applied in one-step procedures.

Although a monograph on the Barbier reaction wouldn't be complete without having mentioned these one-step processes, it is felt that such reactions are not typical Barbier procedures.

Nevertheless some examples of these valuable syntheses will be given here.

On p. 48 of Sect. 2.4.2 mention was made of attack by possible intermediate organomagnesium species on phosphorous in phosphonates, resulting in the formation of cyclic compounds.

Coupling of trimethylchlorosilane and geminal dibromides gave a mixture of products [104]:

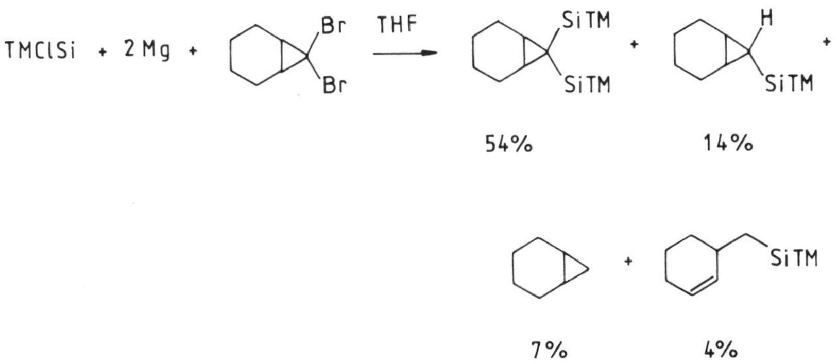

TMClSi + 2 Mg + (Br, Br) —THF→ (SiTM, SiTM) + (H, SiTM) +

54% 14%

+ SiTM

7% 4%

Silacyclobutanes were obtained in high yields when 3-chloropropyl-chlorosilanes were reacted with magnesium [105]; especially noteworthy is the following reaction:

75%

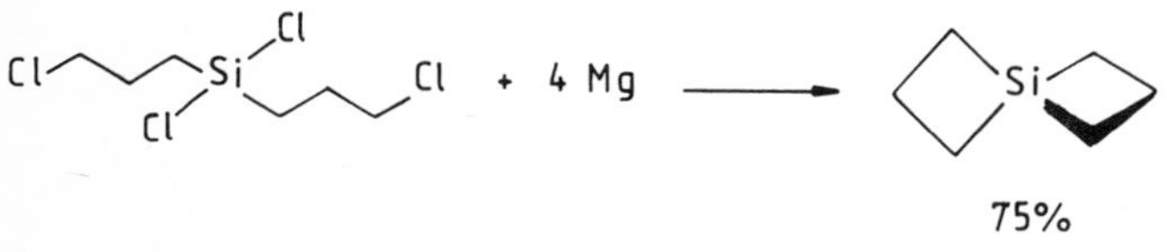

Also aromatic organo halides were used in coupling reactions; in the following example [106] a rather complicated chlorosilane was involved in reactions which gave fair yields:

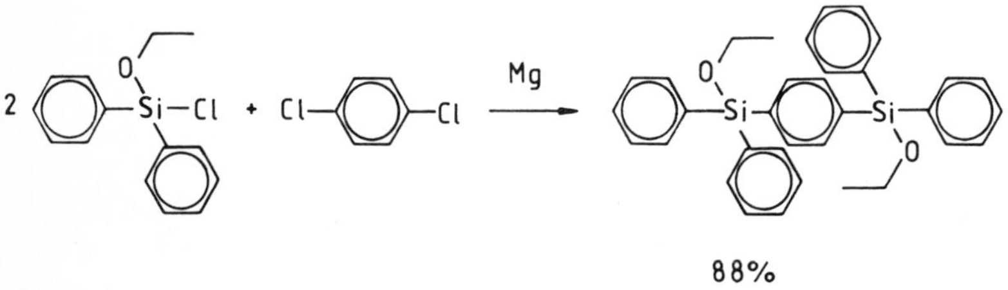

88%

The reaction of the silylating agent TMClSi/Mg/HMPT (named S_1) has been extensively studied over the last three decades [107]. The following reactions [107a] are typical for the synthetic possibilities:

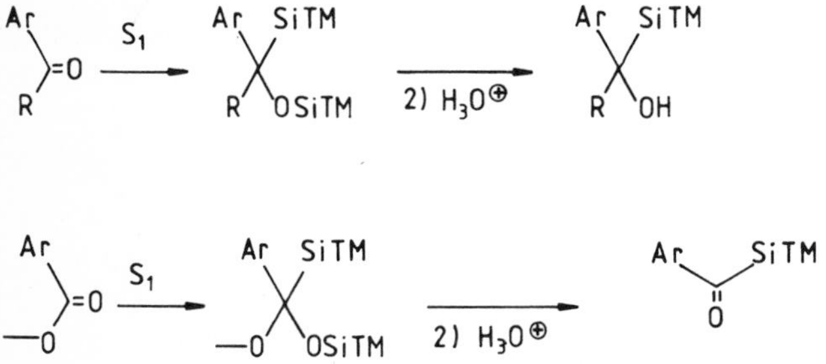

The Grignard reagent obtained from benzyl chloride was reported to give high yields of benzyltrimethylsilane when reacted with TMClSi [108], [109] with some 1,2-diphenylethane; the one-step procedure with benzyl bromide, magnesium and TMClSi in diethyl ether [110] also gave excellent yields of the required coupling product when carried out at $-15°C$

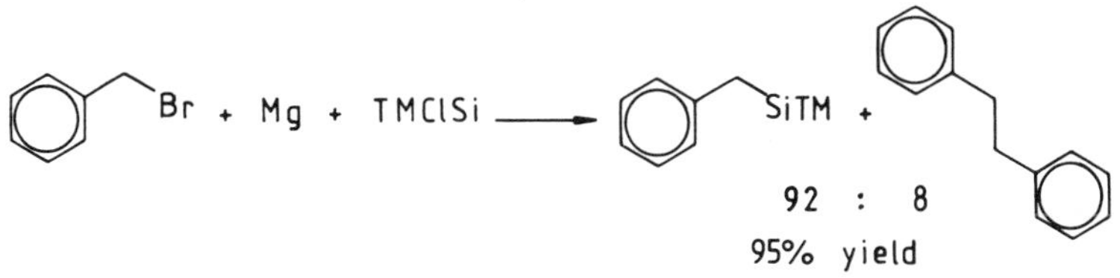

The one-step coupling reaction with metallic halides such as Pb(II) [111], Ge(IV) and several others have found wide application in the synthesis of organometallic compounds. Although the in situ reaction of 1,3-bisbromomethylcyclopentane, magnesium and dichlorosilanes gave satisfactory yields of 3-sila[1,2,3]bicyclooctanes [112], similar reactions with dichlorogermanes resulted in poor yields [113]:

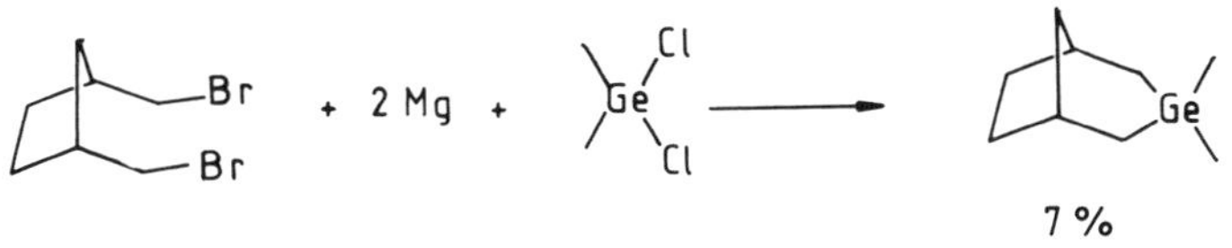

Triorganylboranes were prepared in excellent yields [114] when magnesium reacted with the organo halide in the presence of BF_3-etherate under sonication:

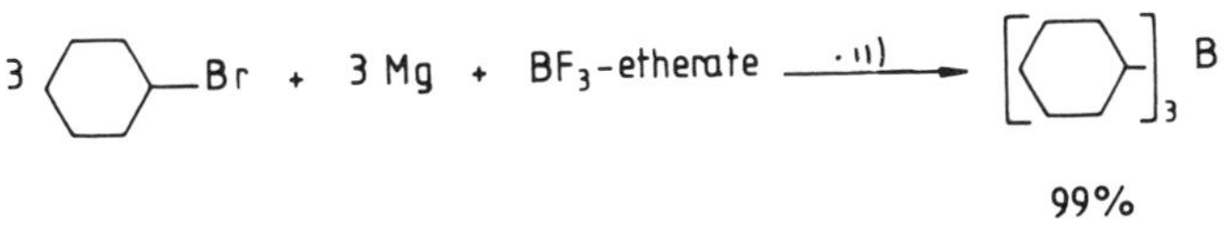

2.5 Conclusion

In this chapter it has been shown that in many instances, results obtained in the one-step procedure, compare very well with those obtained in the classical two-step synthesis.

When it was found, very soon after the discovery of the Grignard reaction (i.e. the Grignard reagent formation reaction), that allylic organomagnesium halides were difficult to prepare, this led to the development of the Jaworsky reaction, the one-step procedure with allylic halides. Yields with this procedure are shown in this chapter to be excellent in spite of original negative reports about it. In view of the reports in this chapter there seems to be no good reason anymore why the two-step procedure should be followed with allylic halides.

After positive results with benzyl chloride in one-step syntheses in the early history of organomagnesium chemistry, Davies and Kipping unambiguously expressed their view, in 1911, that the preparation of the Grignard reagent, was unnecessary and that the one-step process was to be preferred.

Judging by the lack of interest for the Barbier reaction in the period that followed their statement, their voice was apparently not heard.

In 1930, a patent claimed the industrial use of the one-step procedure for syntheses with allylic halides.

It took another thirty years before Dreyfuss convincingly demonstrated that for allylic halides the one-step procedure was the best.

Later, more recent, work demonstrated that also for other halides the one-step procedure may lead to excellent results.

The most favourable reaction conditions have to be found out by experiment: solvent, reaction temperature and the like seem to be more critical in Barbier reactions.

From an economic point of view it seems obvious that the one-step procedure finally will be the most important procedure in organomagnesium chemistry, which – on the whole – is still of great value in synthetic chemistry.

2.6 References

1. Frankland E and Duppa BF (1865) Ann 133: 80
2. Grignard V (1900) Compt Rend 130: 1322
3. Barbier Ph (1899) Compt Rend 128: 110
4a. Reformatsky S (1887) Ber 20: 1210
4b. Reformatsky S and Plesconassoff B (1895) Ber 28: 2838
5a. Zelinsky N and Gutt J (1901) J Russ Phys-Chem Soc 33: 730
5b. Zelinsky N and Gutt J (1902) J Russ Phys-Chem Soc 34: 105
5c. Zelinsky N and Gutt J (1902) Ber 35: 2140
6. von Braun J (1901) Ann 314: 168
7. Schroeter G (1904) Ber 37: 1090
8. Schroeter G (1907) Ber 40: 1589
9. Schroeter G (1908) Ber 41: 5
10. Rupe H and Busolt E (1907) Ber 40: 4537
11. Willstätter R and Hatt D (1919) Ann 418: 148
12. Jacques J and Weidmann Chr, Bull Soc Chim France 1958: 1478
13. Moriwake T (1966) J Org Chem 31: 983
14. Newman MS and Evans Jr EJ (1955) J Amer Chem Soc 77: 946
15. Zeltner J (1908) Ber 41: 589
16. Rassow B and Bauer R (1908) Ber 41: 963
17. Shriner RL (1942) Organic Reactions 1: 1–37
18. Rathke MW (1974) Organic Reactions 22: 423–460
19a. Grignard V (1901) Thèse Univ de Lyon p 23
19b. Grignard V (1901) Ann Chim et Phys Sér 7 24: 450
20. Arbusov A (1901) J Russ Phys-Chem Soc 33: 38; Chem Zentralblatt 1901 I: 998
21a. Jaworsky W (1908) J Russ Phys-Chem Soc 40: 782; Chem Zentralblatt 1908 II: 1412
21b. Jaworsky W (1909) Ber 42: 435

21c. Jaworksy W (1908) J Russ Phys-Chem Soc 40: 1746; Chem Zentralblatt 1909 I: 856
22. Houben J (1903) Ber 36: 2897
23. Lespieau (1907) Compt Rend 144: 1161
24a. Grischkewitsch-Trochimowsky E (1908) J Russ Phys-Chem Soc 40 1685; Chem Zentralblatt 1909 I: 846
24b. Grischkewitsch-Trochimowsky E (1908) J Russ Phys-Chem Soc 40: 761; Chem Zentralblatt 1908 II: 1434
24c. Grischkewitsch-Trochimowsky E (1909) J Russ Phys-Chem Soc 41: 1326; Chem Zentralblatt 1910 I: 739
25. Tarassow B (1909) J Russ Phys-Chem Soc 41: 1309; Chem Zentralblatt 1910 I: 739
26. Kushmin W (1909) J Russ Phys-Chem Soc 41: 1314; Chem Zentralblatt 1910 I: 739
27. Ryshenko P (1909) J Russ Phys-Chem Soc 41: 1695; Chem Zentralblatt 1910 I: 1143
28. Ssemenzow A and Konjuchow-Dobryna P (1911) J Russ Phys-Chem Soc 43: 990; Chem Zentralblatt 1911 II: 1923
29. Mazurewitsch I (1911) J Russ Phys-Chem Soc 43: 973; Chem Zentralblatt 1911 II: 1921
30. Choin M (1912) J Russ Phys-Chem Soc 44: 1844; Chem Zentralblatt 1913 I: 1421
31. Bodroux F and Taboury F (1909) Compt Rend 148: 1675
32. Zelinka Th (1914) Monatsh f Chemie 34: 1507
33. Pariselle H (1912) Compt Rend 154: 710
34. Klimenko D (1912) J Russ Phys-Chem Soc 43: 212; Chem. Zentralblatt 1911 I: 1852
35. Mazurewitsch J (1914) J Russ Phys-Chem Soc 46: 13; Chem. Zentralblatt 1914 I: 1999
36. Koryukin ND (1911) J Russ Phys-Chem Soc 43: 208; Chem Abstr 6: 223 (1912)
37. Orlow A (1912) J Russ Phys-Chem Soc 44: 1868; Chem Zentralblatt 1913 I: 1417
38. Jacubowitsch W (1912) J Russ Phys-Chem Soc 44: 1858; Chem Zentralblatt 1913 I: 1417
39. von Fersen G (1910) J Russ Phys-Chem Soc 42: 681; Chem Zentralblatt 1910 II: 1535
40. Coffey S (1922) Rec Trav Chim Pays-Bas 41: 652
41. Gilman H and McGlumphy JH (1928) Bull Soc Chim France 43: 1322
42. Knorr A (1930) German Patent 544.388; Chem Abstr 26: P2466 (1932)
43. Karrer P, Salomon H, Morf R and Walker O (1932) Helv Chim Acta 15: 878 (1932)
44. Milas NA and McAlevy A (1935) J Amer Chem Soc 57: 580
45. Levina R Ya and Trakhtenberg DM (1936) J Gen Chem USSR 6: 764; Chem Abstr 30: 6338 (1936)
46. Bacon RGR and Farmer EH, J Chem Soc 1937: 1065
47. Henze HR, Allen BB and Leslie WB (1942) J Org Chem 7: 326
48. Fischer FG (1943) Ber 76: 734
49a. Gaylord NG and Becker EI (1950) J Org Chem 15: 305
49b. Dupont G, Dulou R and Christen G, Bull Soc Chim France 1954: 820
50. Davies H and Kipping FJ (1911) J Chem Soc 99: 296
51. Dreyfuss M (1963) J Org Chem 28: 3269
52. Hwa JCH and Sims H (1961) Organic Synthesis 41: 49
53. Hartog FA (1978) Thesis Free University Amsterdam
54. Hayakawa K, Miyazaki O, Nakada K, Kodera K and Ohari K (1990) Jpn. Kokai JP 02 145,547; Chem Abstr 113: P193457 (1990)
55. Young WG and Winstein S (1935) J Amer Chem Soc 57: 2013
56. Young WG, Winstein S and Prater AN (1936) J Amer Chem Soc 58: 289
57. Wilson KW, Roberts JD and Young WG (1950) J Amer Chem Soc 72: 215
58. DeWolfe RH, Roberts JD and Young WG (1957) J Amer Chem Soc 79: 4795
59. Gaudemar M, Bull Soc Chim France 1958: 1475
60. Nordlander JE, Young WG and Roberts JD (1961) J Amer Chem Soc 83: 494
61. Young WG and Roberts JD (1946) J Amer Chem Soc 68: 1472
62. Felkin H and Roussi G, Tetrahedron Lett 1965: 4153
63. Benkeser RA and Broxterman WE (1969) J Amer Chem Soc 91: 5162
64. Felkin H, Frajerman C and Gault Y, J Chem Soc Chem Comm 1966: 75
65. Sakurai H, Kudo Y and Miyoshi H (1976) Bull Chem Soc Japan 49: 1433
66. Richet G and Pecque M (1974) Compt Rend Sér C 278: 1519
67. Zelinsky N and Gutt J (1907) Ber 40: 3049
68. Seetz JWFL, Hartog FA, Böhm HP, Blomberg C, Akkerman OS and Bickelhaupt F (1982) Tetrahedron Lett 23: 1497
69. Bryce-Smith D and Wakefield BJ, Tetrahedron Lett 1964: 3295
70. Bryce-Smith D, Morris PJ and Wakefield BJ, Chem and Ind 1964: 495

71. Koshutin VI (1975) J Org Chem USSR 11: 1775
72. Corey EJ and Kuwajima I (1970) J Amer Chem Soc 92: 395
73. Zelinsky N and Moser A (1902) Ber 35: 2684
74. Leroux Y and Normant H (1967) Compt Rend Sér C 265: 1472
75. Leroux Y, Bull Soc Chim France 1968: 359
76. Danishefski S and Dumas D, J Chem Soc Chem Comm 1968: 1287
77. Mirrington RN and Schmalzl KJ (1972) J Org Chem 37: 2871
78. Crandall JK and Magaha HS (1982) J Org Chem 47: 5368
79. Feugeas Cl, Bull Soc Chim France 1963: 2568
80. House HO and Blaker JW (1958) J Org Chem 23: 334
81. Ginah FO, Donovan Jr Th A, Suchan SD, Pfennin DR and Ebert GW (1990) J Amer Chem Soc 55: 584
82. Larchevêque M, Debal A and Cuvigny Th (1975) J Organometal Chem 87: 25
83. Lambert JB and Oliver WL (1971) Tetrahedron 27: 4245
84. Wetzel RB and Kenyon GL (1972) J Amer Chem Soc 94: 9230
85. Heinicke J, Böhle I and Tzschach A (1986) J Organometal Chem 317: 11
86. Tischtschenko (1887) Journ Russ Phys-Chem Soc 1887 (I) 483; Ber (Ref Pat Nekrol) 20: 704
87. Wagner G and Ginzberg I (1894) Ber 26: 2434
88. Grignard V and Tissier L (1902) Compt Rend 134: 107
89. Fournier H (1907) Comp Rend 144: 331
90. Pariselle H (1909) Compt Rend 148: 849
91. Oddo B and del Rosso G (1911) Gazz Chim Ital 41I: 273; Chem Abstr 5: 2639
92a. Blomberg C and Vreugdenhil AD (1961) Rec Trav Chim Pays-Bas 81: 238
92b. Blomberg C, Vreugdenhil AD, and Homsma TJ (1962) Rec Trav Chim Pays-Bas 82: 355
92c. Westera G (1976) Thesis Free University Amsterdam
92d. Westera G, Blomberg C and Bickelhaupt F (1978) J Organometal Chem 144: 291
93a. Luff BDW and Kipping FS (1908) J Chem Soc 93: 2096
93b. Martin G and Kipping FS (1909) J Chem Soc 95: 302
94a. Chelinzeff W (1906) Ber 39: 779
94b. Chelinzeff W (1907) Ber 40: 1495
95. Schlenk Jr W (1931) Ber 64: 739
96. Miginiac-Groizeleau L (1961) Ann Chim (Paris) 13: 1071
97. Molnarfi S (1964) Fr Patent 1964: 1.359.195; Chem Abstr 61: P8236g (1964)
98. Bertini F, Grasselli P, Zubiani G and Cainelli G (1970) Tetrahedron 26: 1281
99. Obe Y, Sato M and Matsuda Ts (1971) Kyushu Daryaku Kogaku Shuhu 44: 208; Chem Abstr 77: 48546z (1972)
100. Kato K and Hirayama Y (1973) Japanese Patent 7430333; Chem Abstr 81: 135694x (1974)
101. Stock O, Troupel M and Périchon J (1985) Tetrahedron Lett 26: 1509
102. Sibille S, d'Incan E Report L and Périchon J (1986) Tetrahedron Lett 27: 3129
103. Ashby EC and Al-Fekri DM (1990) J Organometal Chem 390: 275
104. Seyferth D and Lambert Jr RL (1975) J Organometal Chem 88: 287
105. Damrauer R, Davies RA, Burke MT, Karn RA and Goodman GT (1972) J Organometal Chem 43: 121
106. Bazant V and Cerny M (1974) Collect Czech Chem Commun 39: 1728, 1735 and 1880
107a. Calas R (1980) J Organometal Chem 200: 11
107b. Dunoguès J, Chem tech 1982: 373
107c. Dunoguès J, Bull Soc Chim France 1987: 659
108. Gilman H and Marshall FJ (1949) J Amer Chem Soc 71: 2066
109. Hauser CR and Hance CR (1951) J Amer Chem Soc 73: 5846
110. Muzart J and Riahi A (1991) Synth Comm 21: 1247
111. Williams KC and Cook SE (1973) US Patent 3,775,454; Chem Abstr 80: 37289 (1974)
112. Blankenship C and Cremer S (1986) Organometallics 5: 1329
113. Sommer AG, Cremer SE, Campbell JA and Thompson MR (1990) Organometallics 9: 1784
114. Jadhav PK, Bhat KS, Perumal PT and Brown HC (1986) J Org Chem 51: 432

3 Barbier-Type Reactions with Other Metals

3.1 Introduction

Since the 1970s the use of metals, other than magnesium, for one-step synthetic processes has gained tremendous momentum.

On the one hand, the more electropositive lithium is enjoying increasing 'popularity' in this field; particularly in synthetic applications with sterically crowded reagents.

On the other hand, less electropositive metals such as zinc, aluminium and tin have drawn the attention of synthetic chemists in their search for new and more efficient routes to products.

As will be illustrated in this chapter, these new developments have been so extensive that it is practically impossible to keep track of all the important new publications.

It is inevitable that in the near future monographs on the synthetic applications of several of the individual metals will appear since each one of them may have its own advantages in synthesis.

Aside from new metals, novel technical and practical developments have widened the scope of the one-step procedure. Examples of these are the introduction of highly reactive metals (either through their fine particle size or through their graphite intercallation compounds) or the use of sonochemical catalysis.

As for the names of these reactions with metals, other than magnesium, it has already been stated in Chapter One that preference will be given to the name *Barbier-type reaction* instead of simply *Barbier reaction*. Furthermore the name of each individual metal will be included in the indication of the processes; hence there will be mention of *Li-Barbier reactions*, *Zn-Barbier reactions* and so on.

3.2 Sodium in One-step Processes

Although the number of applications of sodium in one-step processes is limited, some Na-Barbier reactions are worth mentioning here.

The metal was the first to be generally applied in organic synthesis. In 1855, after initial attempts with potassium one year earlier [1], Adolphe Wurtz [2], reacted sodium with haloalkanes. This attempt to produce *radicals* was similar to Frankland trials with zinc a few years earlier (Sect. 1.2), and introduced the longest known *name-reaction* in organic chemistry, the Wurtz reaction:

2 haloalkane + 2 Na ⟶ alkyl-alkyl + 2 Na-halide

In 1864, Tollens together with Fittig [3], widened the scope of the Wurtz reaction by applying it to a mixture of aliphatic and aromatic organohalides, thus introducing the Wurtz-Fittig reaction.

One year later an unsuccessful attempt to couple two benzoyl groups (from benzoyl chloride) with sodium was reported [4]. This reaction was later carried out in 1883 [5].

In 1895, four years before Barbier's report [6] on the one-step synthesis with magnesium, sodium was used in a *one-pot* reaction with bromobenzene and ethyl oxalate, the unforeseen product being triphenylmethanol [7].

Attempts to use sodium in a coupling reaction with benzoyl chloride and a haloalkane [8], as tried earlier by Freund [9] in 1861 (Sect. 1.2.3) were reported in 1907.

After the introduction of magnesium in synthetic organic chemistry at the beginning of this century [10, 11] interest in sodium in this type of reaction faded.

In 1968, sodium was successfully applied [12] for the following cyclization reaction (intramolecular Barbier-type reaction) which had previously failed with magnesium:

O= … Br —— Na/THF, 2) $H_3O^{\oplus}$ ⟶ HO— … H

A few years later the same cyclization reaction was successfully performed with the iodo- instead of the bromoketone [13].

It was in that period that lithium was successfully introduced in what was named *A One-step Alternative to the Grignard Reaction* (Sect. 3.3) [14].

In 1972, in a follow-up to the first paper on this Li-Barbier reaction [15], also the use of sodium in this type of reaction was studied. It was reported that there was a significant diminution in the yields of alcohols; that was particularly true of carbonyl compounds which possessed α-hydrogen atoms.

Recently sodium was used in a Barbier-type reaction [16] with bromobenzene and *tert*-butyl isocyanate in THF and a stoichiometric amount of HMPA (using sonication,·))), to promote the reaction):

Br + Na + N=C=O THF/HMPA 1) → [Na⊕ ⊖ N] I → 2) $H_3O^{\oplus}$ → 78%

Lithiation of I with *n*-butyllithium followed by reaction with an electrophile such as dimethylformamide yielded *ortho*-substituted *tert*-butylbenzamides in fair yields:

I + *n*-butylLi 2) DMF 3) $H_3O^{\oplus}$ → N OH 67%

3.3 Lithium in Barbier-Type Reactions

3.3.1 Cyclization Reaction; 1968

At the end of the 1960s interest grew in the use of lithium in Barbier-type reactions.

In a study of cyclization reaction of haloketones with lithium [17] two different types of reaction were shown to take place:

a. in a strongly coordinating solvent – such as THF/HMPA – 5-bromopentan-2-one gave almost exclusively methyl cyclopropyl ketone:

Br + Li THF/HMPA 2) $H_3O^{\oplus}$ →

b. in diethyl ether, at − 10 °C, followed by warming up of the reaction mixture to + 10 °C, 1-methyl-1-cyclobutanol was formed in 30% yield:

Br + Li Et_2O; -10 °C 2) $H_3O^{\oplus}$ → OH 30%

In a mechanistic study on homo-enolate anions [18] lithium was reacted with a β-halogenated aldehyde or ketone followed by acetylation of the reaction mixture.

The cyclopropyl acetates resulting from an intramolecular Barbier-type of reaction were obtained in rather satisfactory yields:

Br, =O, H + Li 2) Ac_2O → OAc 40-60%

Br, =O + Li 2) Ac_2O → OAc 49%

3.3.2 One-Step Alternative to the Grignard Reaction

A real breakthrough in the study of Barbier-type reactions with metals different from magnesium came with the publication of *A one step Alternative to the Grignard Reaction* [14].

Fair to excellent yields were reported from Li-Barbier reactions with a variety of organic halides and aldehydes, ketones and esters (Table 3.1).

Results from a later publication [15] are presented in Tables 3.2 and 3.3.

The following comments were made:

1. Organic bromides seem to give the best results. The lower yields with iodides are *almost certainly due to increased probability of a Wurtz condensation reaction taking place between the organolithium species and the iodide.*
2. Both alkyl and aryl halides give high yields.
3. Secondary halides (cyclopentyl and -hexyl) may be employed with similar ease.
4. The size of the halide or of the aldehyde does not influence the yields significantly.
5. The low yield obtained with propanone is surprising and is probably due to pinacol formation.

In order to ascertain the sensitivity of the process to changes in the experimental parameters, the reaction of 1-bromopropane with butanal in the presence of lithium was studied in some detail.

Two remarks were made regarding the scope and limitations of this elegant one-step process.

a. When the reaction is carried out with α,β-unsaturated carbonyl compounds, addition takes place almost exclusively to form 1,2-adducts. Magnesium organic compounds mainly give 1,4-addition reaction products.

Table 3.1. Yields of alcohols from Barbier-type reactions of organic halides with lithium and a substrate in THF as the solvent [14]

Substrate	Halide	Yield (%)
$HC(O)CH_2CH_3$	CH_3CH_2Br	90
$HC(O)CH_2CH_3$	CH_3I	51
$CH_3C(O)CH=C(CH_3)_2$	C_6H_5Br	67
$(HC(O)H)_n$	$Br(CH_2)_5CH_3$	72
$CH_3C(O)CH_3$	C_6H_5Br	34
$C_6H_5C(O)OCH_3$	CH_3I	74
$C_6H_5C(O)H$	C_6H_5Br	96
"	$CH_2=CHCH_2Br$	71
$CH_2=C(CH_3)C(O)OCH_3$	$CH_3(CH_2)_3Br$	89

b. Benzyl halide cannot be efficiently used in this type of reaction. Benzyllithium itself is difficult to prepare because of its marked tendency to undergo Wurtz coupling. Indeed large quantities of 1,2-diphenylethane were found among the reaction products in all cases.

One way to overcome this problem was to use benzyl methyl ether as is illustrated in the following example:

$$C_6H_5CH_2OCH_3 + Li + CH_3C(O)CH=C(CH_3)_2 \xrightarrow[2)\ H_3O^{\oplus}]{THF} C_6H_5CH_2C(CH_3)(OH)CH=C(CH_3)_2$$

Table 3.2. Yields of alcohols from Barbier-type reactions of organic halides with lithium and aldehydes in THF as the solvent [15]

Substrate	Halide	Yield (%)
$(HCHO)_n$	$Br(CH_2)_5CH_3$	72
,,	$Br(CH_2)_9CH_3$	80
$HCOCH_3$	$Br(CH_2)_4CH_3$	39
$HCOCH_2CH_3$	ICH_2CH_3	51
,,	$BrCH_2CH_3$	90
$HCO(CH_2)_2CH_3$	$BrCH_2CH_2CH_3$	89
,,	$Br(CH_2)_3CH_3$	73
,,	$Br(CH_2)_4CH_3$	73
,,	$BrCH_2CH(CH_3)_2$	89
,,	$BrC(CH_3)_3$	89
$HCOCH(CH_3)_2$	$Br(CH_2)_3CH_3$	89
$HCO(CH_2)_4CH_3$	$BrCH_3$	70
$HCO\text{-}n\text{-}C_9H_{19}$	$Br\text{-}n\text{-}C_{10}H_{21}$	75
,,	$Br\text{-}n\text{-}C_{18}H_{37}$	80
$HCO\text{-}n\text{-}C_{12}H_{25}$	$Br\text{-}n\text{-}C_{16}H_{33}$	50

Table 3.2 (continued)

Substrate	Halide	Yield (%)
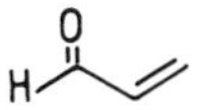 O H	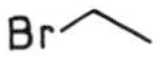 Br	90
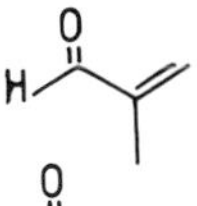 O H	 Br	93
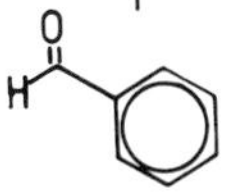 O H	 Br	71
''	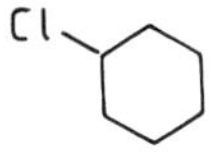 Cl	60
''	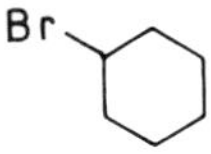 Br	72
''	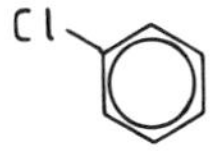 Cl	100
''	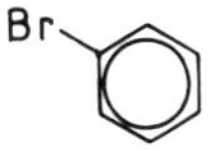 Br	94
''	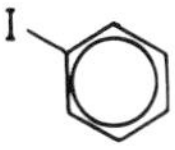 I	64

Table 3.3. Yields of alcohols from Barbier-type reactions of organic halides with lithium and different substrates in THF as the solvent [15]

Substrate	Halide	Yield (%)
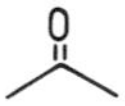 O	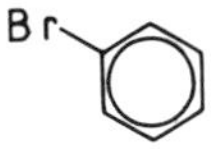 Br	34
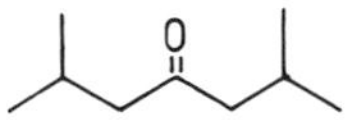 O	 Br	90
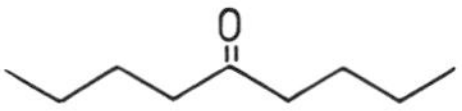 O	''	91
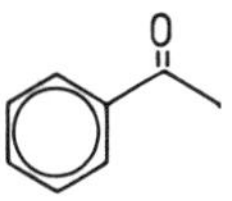 O	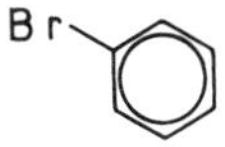 Br	62

Table 3.3 (continued)

Substrate	Halide	Yield (%)
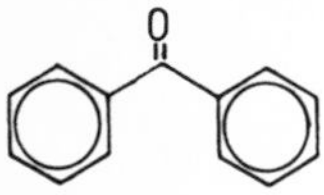 O	 Br	95
''	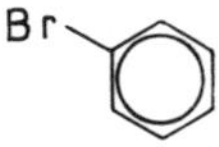 Br	90
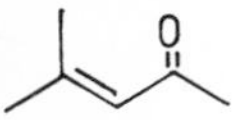 O	''	68*
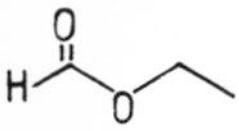 O H O	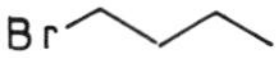 Br	91
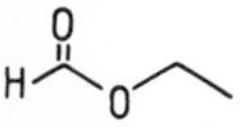 O H O	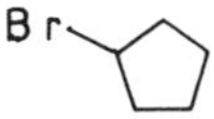 Br	85
''	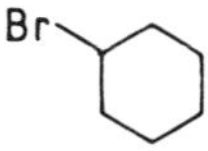 Br	80
''	$Br-n-C_{10}H_{21}$	62
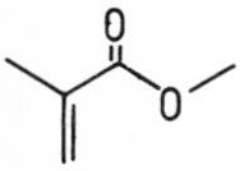 O O	 Br	89
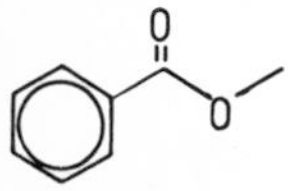 O O	I	74

* Two products were identified: 32% of 2-methyl-4-phenylpenta-1, 3-diene and 36% of 4-methyl-2-phenylpenta-1, 3-diene, which, to the authors' opinion, result from alternative dehydration routes from the 1, 2-adduct, 4-methyl-2-phenyl pent-3-en-2-ol, rathr than from any 1,4-adduct present

A second way to use benzyllithium in this one-step process is to form a lithium/electron-acceptor complex in THF before addition of the mixture of reactants. For example, lithium can be dissolved in a naphthalene solution in THF to form a dark green solution of lithium/naphthalene. When the halide-carbonyl mixture was added to an excess of this complex an extremely fast electron-transfer reaction occurred, which minimised the stationary concentration of benzyl halide and thereby lowered the probability of undergoing Wurtz condensation.

The reaction of benzaldehyde with benzyl bromide in the presence of an excess of lithium/naphthalene gave a 74% yield of 1,2-diphenylethanol at 0 °C:

Li/naphthalene in THF at 0 °C; 2) $H_3O^{\oplus}$; 74%

Although the mechanistic details of the Li-Barbier reaction will be discussed in Sect. 4.4.2, it should be mentioned here that a mechanism in which an intermediately formed radical anion (or ketyl) reacts with the organic halide to give an alkoxide radical as given in the following equations:

$$\text{>C=O} + \text{Li} \longrightarrow \text{>}\dot{\text{C}}\text{-O-Li} \xrightarrow{\text{RX}} \text{LiX} + \text{R-C-O}\cdot \xrightarrow{\text{Li}} \text{R-C-OLi}$$

had to be rejected since the blue colour characteristic of the ketyl benzophenone/lithium in THF was not discharged after a stoichiometric quantity of *n*-butyl bromide had been added.

The following conclusion was therefore made:

Thus it appears that intermediate radical-anion formation does not occur in the pseudo-Grignard process and that, as expected, the reaction proceeds via the formation of organolithium compounds.

Li-Barbier reactions could be used to synthesize aldehydes and ketones when carboxamides or carbamates were applied as the substrates [19].

The following equation illustrates the process with carboxamides:

$$\text{R-Br} + \text{Li} + \text{R'C(=O)NR''}_2 \xrightarrow[\text{2) } H_3O^{\oplus}]{} \text{RC(=O)R'}$$

Table 3.4 lists the results; it is apparent from these data that aromatic amides react smoothly to give ketones in good yield but the reaction with aliphatic amides is not an efficient process.

Table 3.4. Yields of aldehydes and ketones from Li-Barbier reactions of organic bromides and carboxamides [19]

Organic bromide	Carboxamide	Product	Yield (%)
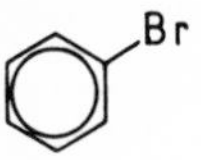 Br	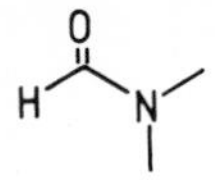 O H N	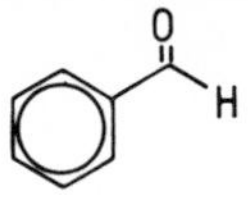 O H	78
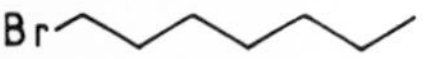 Br	"	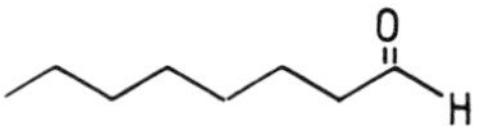 O H	56
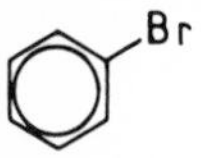 Br	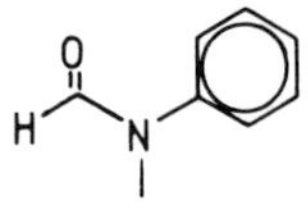 O H N	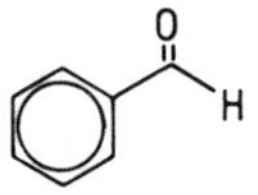 O H	56
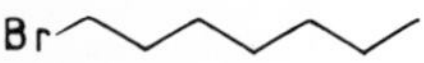 Br	"	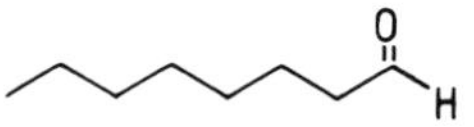 O H	73
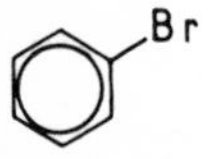 Br	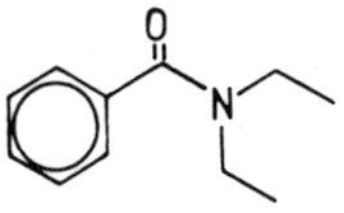 O N	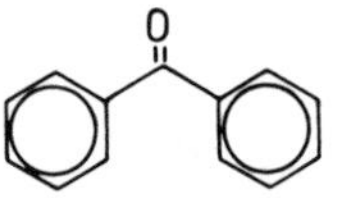 O	76
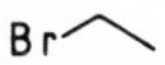 Br	"	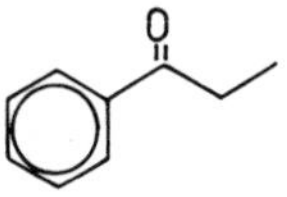 O	79
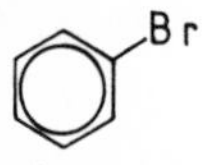 Br	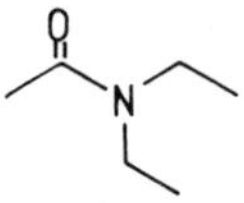 O N	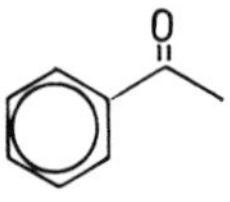 O	11
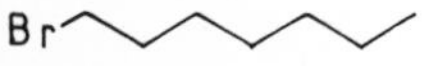 Br	"	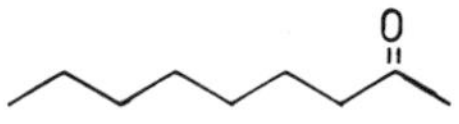 O	10
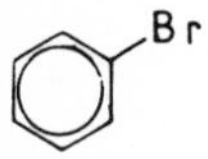 Br	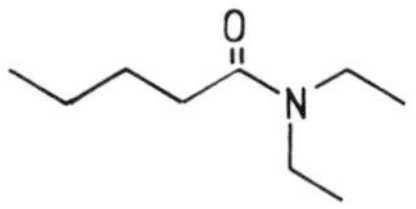 O N	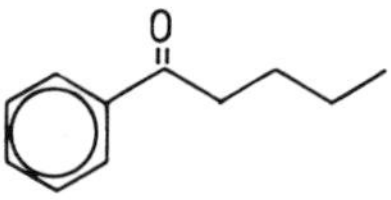 O	23

The reaction of bromobenzene with ethyl *N*, *N*-diethylcarbamate leads to the formation of benzophenone in high yields:

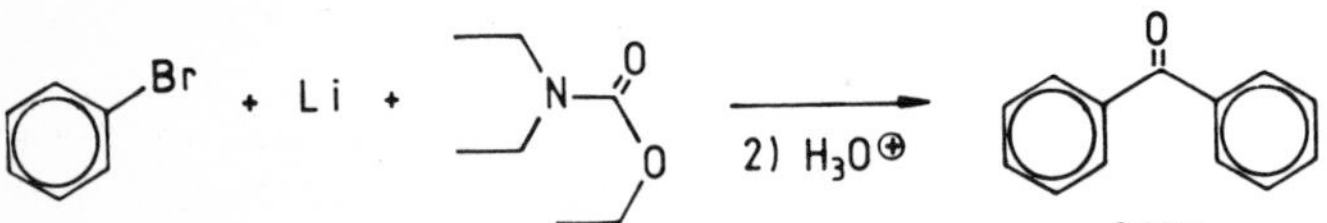

Br + Li + N O O → 2) $H_3O^{\oplus}$ O

82%

With *n*-bromoheptane, however, only 23% of the expected ketone was formed. As the major product ($\pm$ 70%) *N*, *N*-diethyloctamide was formed which the author considered unexpected

since it would seem reasonable to assume that amides might be more reactive towards organo-metallic derivatives than carbamates.

3.3.3 Allylic Mesitoates

To synthesize 1,5-dienes via cross-coupling reactions, the in situ reactions were studied of allylic mesitoates, allylic bromides and lithium in tetrahydrofuran at 0 °C [20, 21]:

Li, THF; 0 °C

Substituents R as given above may vary from H and methyl to 4-methylpent-3-en-1-yl and 4,8-dimethylnona-3,7-dien-1-yl. Yields of such couplings could be as high as 95%.

It was also found that the in situ generation of allylic organolithium compounds (this method avoids the necessity of synthesizing an allylic halide) in this manner can be used to produce allylic carbinols when aldehydes and ketones were present in the reaction mixture as electrophiles.

3.3.4 Cyclization Reactions; 1975

The results of intramolecular Li-Barbier reactions (cyclization reactions) with 5-bromo-1-cyano-2,2-dimethylpentane and the corresponding 6-bromohexane were rather disappointing [22]. The expected 2,2-dimethylhexanone was formed in only 31% yield:

Br–(CH₂)₃–C(CH₃)₂–CN + Li → (Et_2O / HMPA; 2) $H_3O^{\oplus}$) 31%

The formation of cycloheptanone occurred in even lower yield.

It is worth mentioning that the substituted cyclohexanone was also formed from the iodocyanopentane when lithiated with *n*-butyllithium in diethyl ether. The halogen exchange reaction is obviously much faster than nitrile addition for the organolithium reagent (see also the reaction mentioned on p.96.

I–(CH₂)₄–C(CH₃)₂–CN + *n*-BuLi —(Et_2O; 2) $H_3O^{\oplus}$)→ 2,2-dimethylcyclohexanone

3.3.5 Reactions with Nitriles and Nitrosoalkanes

The results of Li-Barbier reactions of bromobenzene and *n*-bromobutane with nitriles have also been compared with those from two-step procedures involving Grignard and organolithium reagents [23].

PhBr + Li + R-CN —(2) $H_3O^{\oplus}$)→ PhC(O)R

The data from Table 3.5 demonstrate that the one-step procedure is *not* preferred for the preparation of these ketones.

However, the results, listed in Table 3.6, show that the Li-Barbier reaction of (substituted) bromobenzenes and 2-methyl-2-nitrosopropanes, which leads to the formation of (substituted) aryl-*tert*-butylhydroxylamines (isolated as nitroxides because of spontaneous oxidation on exposure to air), is preferred over the two-step procedure with Grignard reagents.

Ar-Br + Li + t-Bu-NO —(2) $H_3O^{\oplus}$)→ t-Bu-N(Ar)OH

3.3.6 Reactions with Sterically Crowded Halides; 1978

The discovery, at the end of the 1970s [24], of a remarkably unreactive solution of a tertiary organolithium compound was the start of an interesting expansion of Li-Barbier chemistry.

Table 3.5. Yields of ketones from reaction of various nitriles with organobromides and lithium [23]

Organobromide	Nitrile	Product	Yield (%)	Yield from reaction with	
				RMgX (%)	RLi (%)
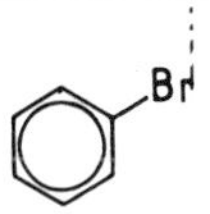	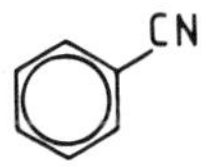	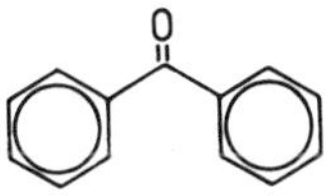	64	65	23
″	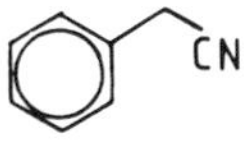	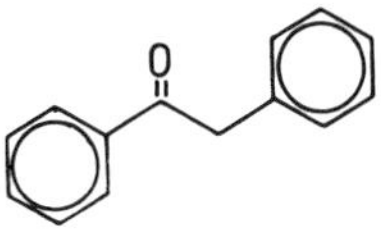	3	33	
″	—CN	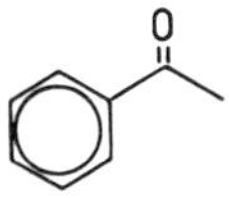	1	70	36
″	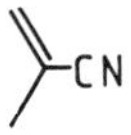	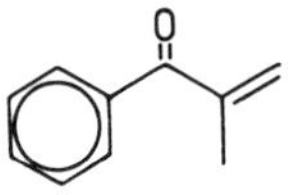	2		
″	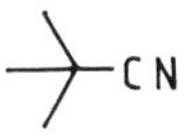	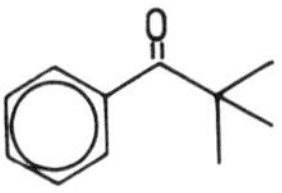	59	100	
	″	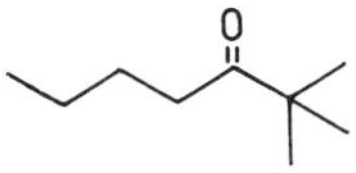	27		
″	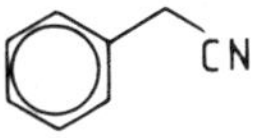	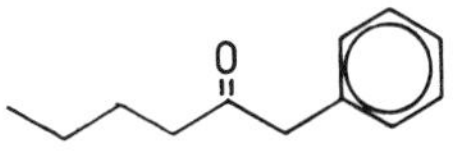	6.5		49

1-Lithioadamantane, formed by reaction of 1-bromo- or 1-chloroadamantane (1-AdBr or 1-AdCl) with lithium in diethyl ether at $-20\,°C$, yielded 75% 1-Ad-D on deuterolysis. Addition of hexamethylacetone (HMA) to the reaction mixture yielded virtually no carbonyl addition reaction product.

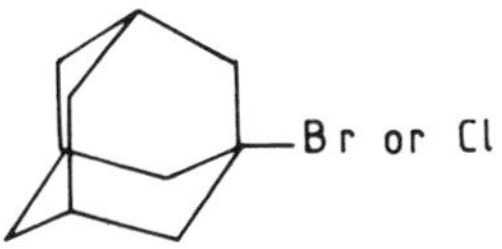

1-Ad.Br or -Cl

Table 3.6. Nitroxide yields from reaction of aromatic bromides with 2-methyl-2-nitrosopropane and lithium [23]

Aryl bromide	Nitroxide	Yield (%)	Yield (%) from RMgX reaction
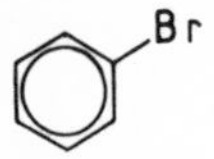	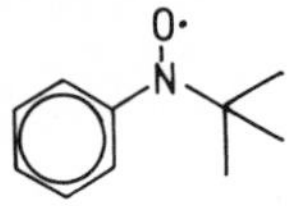	trace	56
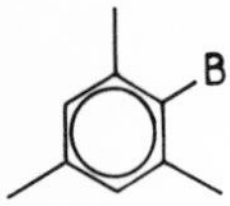	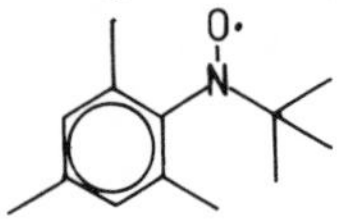	44	32
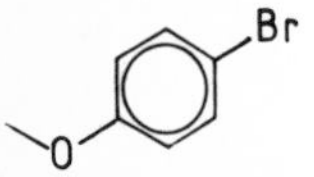	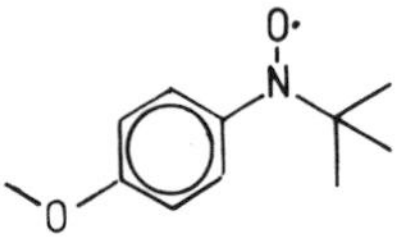	62	
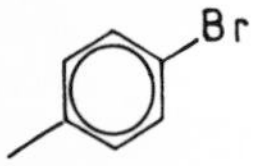	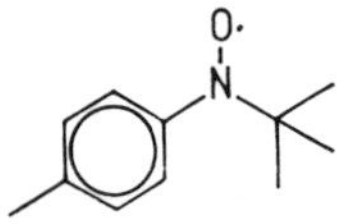	37	38
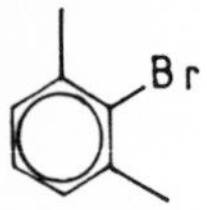	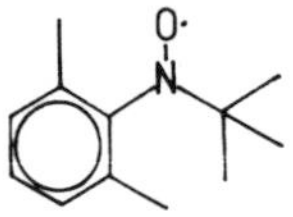	71	52

In contrast, when this reaction was performed as a one-step process, on deuterolysis, products were obtained which suggested that an intermediate organolithium compound had been formed:

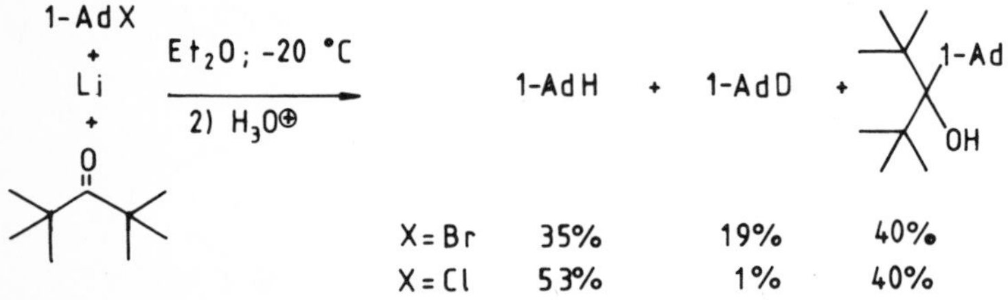

1-AdX + Li + (t-Bu)₂C=O → [Et_2O; −20 °C; 2) $H_3O^{\oplus}$] 1-AdH + 1-AdD + (t-Bu)₂C(1-Ad)OH

	1-AdH	1-AdD	alcohol
X = Br	35%	19%	40%
X = Cl	53%	1%	40%

The mechanistic aspects of these results will be discussed in Sect. 4.4.

Li-Barbier reactions were studied [25] for several other 'cage-structure' bromides with adamantone, another highly sterically hindered reactant, as the carbonyl compound.

The results are presented in Table 3.7.

Comparing the yields of the different products in these reactions with the stability of the cage radicals, the conclusion was drawn that the Barbier-type

Table 3.7. Results of Barbier-type reactions of cage-structure bromides with adamantone and lithium in diethyl ether at $-20\,^{\circ}C$ [25]

Products	RBr, %				
	 Br	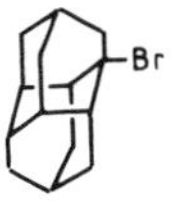 Br	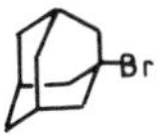 Br	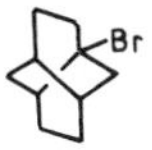 Br	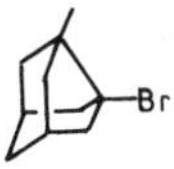 Br
Temperature: 20 °C					
alcohol[a, b]	0	28	71	84	72
pinacol[a]	75	40	29	16	28
RLi[b]	0	0	1	3.5	5.5
hydrocarbons[b]	100	65	27.5	12.5	22.5
Temperature: 45 °C					
alcohol[a, b]	0	36	71	66	20
pinacol[a]	75	37	29	34	80
RLi[b]	0	0	6	18	51
hydrocarbons[b]	100	65	23	16	29

[a] Yields calculated with respect to the ketone
[b] Yields calculated with respect to the halide

reaction could occur without the in situ formation of the organometallic compound (see also Sect. 4.4). Furthermore the organometallic pathway – as opposed to the radical pathway in which a ketyl radical reacts with the transitory species, R · -X- or R · · Li, formed by a single electron transfer from lithium to the organohalide – is favoured when the lifespan of the cage radical decreases (see also p. 154 for the intermediacy of radical-anions in Li-Barbier reactions).

The efficiency of the Li-Barbier reaction for the synthesis of highly sterically hindered tertiary alcohols was further demonstrated when eleven highly sterically hindered, tertiary alcohols were synthesized [26] via this one-step procedure using the following halides

1-bromoadamantane (Br-Ad),
1-bromobicyclo[2.2.1]heptane (Br-Hept) and
1-bromobicyclo[2.2.2]octane (Br-Oct).

The ketones applied were

di(1-bicyclo[2.2.2]octyl)ketone ((Oct)(Oct)ketone),
di(1-adamantyl)ketone ((Ad)(Ad)ketone),
di(1-bicyclo[2.2.1]heptyl)ketone ((Hept)(Hept)ketone),

(Oct)(Hept)ketone,
(Oct)(Ad)ketone,
(Hept)(Ad)ketone and
tert-butyl(Hept)ketone.

The following results were obtained:

lithium with Br-Ad and (Ad)(Nor)ketone	62%
lithium with Br-Ad and (Nor)(Nor)ketone	78%
lithium with Br-Oct and *t*-butyl(Nor)ketone	81%
lithium with Br-Oct and (Oct)(Oct)ketone	56%
lithium with Br-Oct and (Nor)(Nor)ketone	93%
lithium with Br-Nor and (Ad)(Oct)ketone	88%
lithium with Br-Nor and (Oct)(Oct)ketone	62%
lithium with Br-Nor and (Nor)(Nor)ketone	78%

Only when the product alcohol was very congested did the yields drop significantly. For example as in the following combinations of reactants:

lithium with (Br-Ad) and (Ad)(Ad)ketone:	0%
lithium with (Br-Ad) and (Ad)(Oct)ketone:	29%
lithium with (Br-Ad) and (Oct)(Oct)ketone:	40%

The preparation of tri-1-adamantylmethanol, which failed via the one-step Barbier-type reaction with Br-Ad and (Ad)(Ad)ketone (vide supra) was brought about by the one-step reaction of 2.8 equivalents of Br-Ad with one equivalent of the methyl ester of 1-adamantylcarboxylic acid; yield 27%.

3.3.7 Ultrasound in Li-Barbier Reactions

A very powerful new tool was applied in synthetic chemistry with ultrosonic irradiation of reaction mixtures [27, 28]. It is symbolized with the following icon ·)) above or below the reaction arrow.

Sonication was successfully introduced in organometallic chemistry in 1980 [29] when it was applied to Li-Barbier reactions. In most cases these were completed in 10–15 minutes. Normally observed sidereactions such as enolization and reduction seemed to have been minimised.

Table 3.8 presents the results obtained.

Two extraordinary observations require special mention:

1. This reaction could be performed in wet technical grade solvent at room temperature although more of the halide then had to be used, presumably due to hydrolysis of intermediate organolithium compounds.
2. The use of ultrasound allows the direct in situ formation of benzyllithium from benzyl bromide while normally Wurtz coupling predominates with this halide. (See e.g. p. 79 [15]). Condensation, e.g. with acetophenone gives the expected product in 95% yield. Also allylic and vinylic halides undergo clean

Table 3.8. Yields of Li-Barbier reactions under ultrasonic irradiation [29]

Organohalide	Carbonyl Compound	Product	Yield (%)
I	O O H	O OH	100
Br	O H	OH	100
Br	O O H	O OH	100
Br	O	OH	100
Br	O	OH	80
Br	O	HO	100
″	O	OH	84
Br	″	OH	76
Br	O	OH	100
Br	O	HO	95
Br	″	HO	92

Table 3.8 (continued)

Organohalide	Carbonyl Compound	Product	Yield (%)
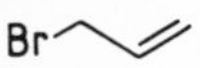 Br	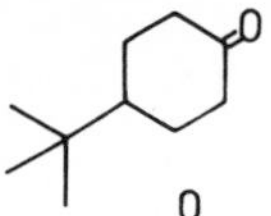 O	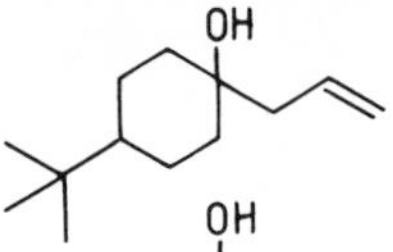 OH	76
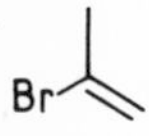 Br	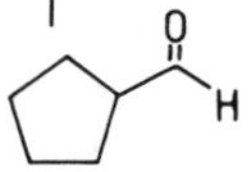 O, H	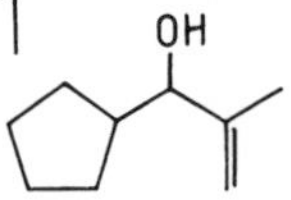 OH	96
Br	"	OH	95

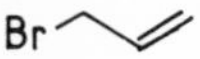

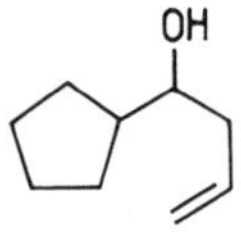

and high-yield reactions under the ultrasound Barbier-type reaction conditions, whereas their transformation to the corresponding lithio derivatives is usually difficult.

Several other sonochemically catalyzed Li-Barbier reactions were published shortly after this first one.

The following cyclization reaction gave excellent yields of the expected product when sonication was applied [30]:

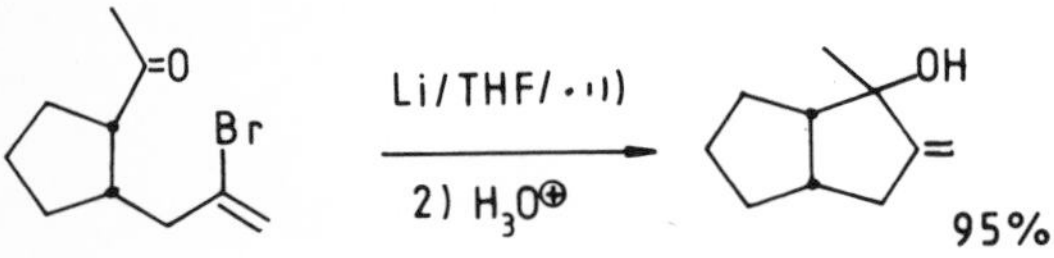

=O Br
Li/THF/·))
2) $H_3O^{\oplus}$
OH
95%

Conventional methods of organometallic generation with lithium and magnesium had failed to give satisfactory results.

The use of low-temperature halogen-metal exchange reaction (*tert*-BuLi at − 20 °C) also did not yield any closure products.

Another similar cyclization reaction [31] that was successfully performed was the following:

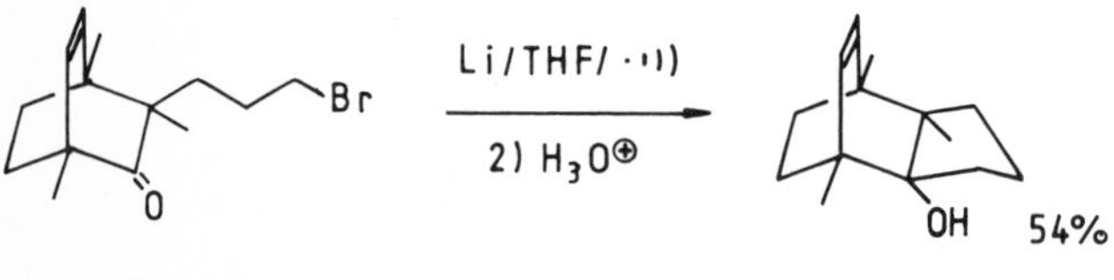

Br
O
Li/THF/·))
2) $H_3O^{\oplus}$
OH 54%

An ultrasound promoted Li-Barbier reaction was used to selectively prepare the 1,2-addition products in the following reaction [32]:

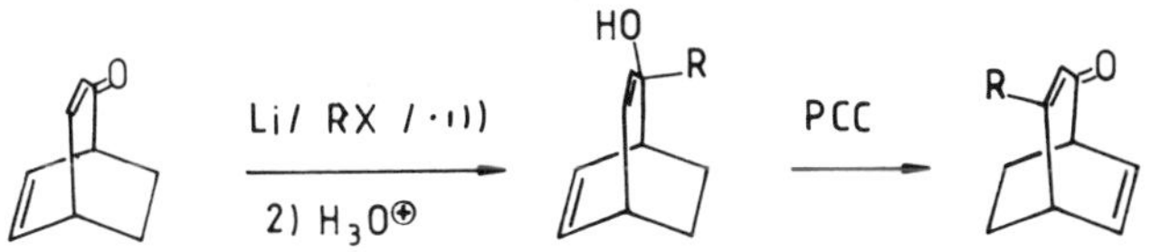

Li/ RX /·)))
2) $H_3O^{\oplus}$
HO R
PCC
R O

Oxidation of the tertiary allylic alcohol with pyridinium chlorochromate (PCC) gave the unsaturated ketone.

With CH_3I as the organo halide the overall yield was 78%; i-C_3H_7Br yielded 90% of the intermediate allylic alcohol, and the α, β-unsaturated ketone in 88%.

For bromobenzene and 2-bromopropene the results were less satisfactory: 33% and 59% yields respectively.

A fair yield was reported from the reaction of 4,4-dimethyl-cyclopent-2-enone with 5-bromo-1-hexene [33]:

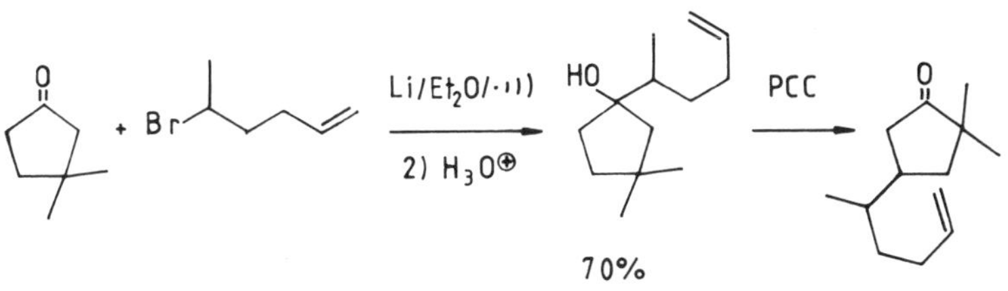

O + Br
Li/Et_2O/·)))
2) $H_3O^{\oplus}$
HO
70%
PCC
O

The stereochemical outcome of the ultrasound promoted Li-Barbier reaction of S(+) 2-halooctanes and cyclohexanone was studied [34] under various reaction conditions (temperature and ultrasonic energy):

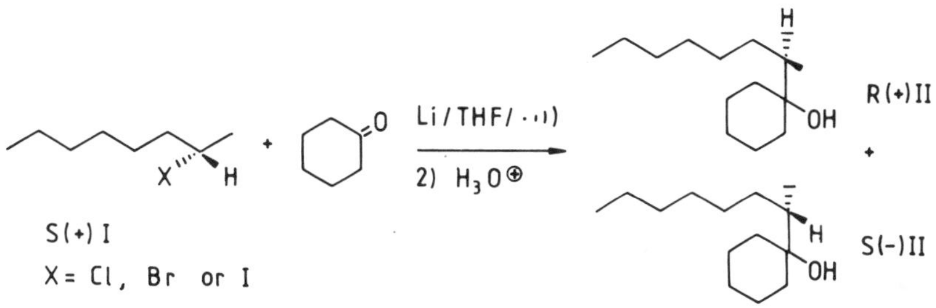

X H
S(+) I
X = Cl, Br or I
+ O
Li/THF/·)))
2) $H_3O^{\oplus}$
H OH R(+)II
+
H OH S(−)II

The enantiomeric excess as well as the absolute configuration of the products (II) suggested the existence of two reactive intermediates following different stereochemical pathways to the end product. The mechanistic implications of these results will be discussed in Sect. 4.4.

A recent study [35] of the fundamental aspects of the sonochemical Barbier reaction was carried out with 1-bromoheptane and benzaldehyde in THF under various reaction conditions (among which were reaction temperature from $-50\,°-+20\,°C$, and intensity of the ultrasonic waves, expressed in the voltage applied to the sonicator).

$$PhCHO + Li + n\text{-}C_7H_{15}Br \xrightarrow[2)\ H_3O^{\oplus}]{THF/\,\cdot\,))} n\text{-}C_7H_{15}CH(OH)Ph$$

Observed yields were as high as 95%. The major sidereaction is the formation of 1,2-diphenyl-1,2-ethanol:

$$2\,PhCHO + 2\,Li \xrightarrow{\cdot\,))} 2\left[Ph\dot{C}H(O^{\ominus}Li^{\oplus})\right] \xrightarrow{2)\ H_3O^{\oplus}} PhCH(OH)CH(OH)Ph$$

This was formed in 13% yield when the reaction was carried out at $-50\,°C$ with a relatively low voltage for the ultrasound generator. Under other reaction conditions the pinacol formation was of the order of 3% (see also Sect. 4.4 for a discussion of the mechanism of this reaction, and Sect. 5.4.2 for some technical details of it).

3.3.8 Other Li-Barbier Reactions with Sterically Hindered Reagents; 1988

As shown earlier, Li-Barbier reactions are very effective in the synthesis of sterically crowded alcohols (pp. 86–90).

However, little success was met in the one-step synthesis of 1-(2-adamantyl)-2-methyl-2-phenyl-1-ethanol from 2-chloro-2-phenylpropane, adamantanone and lithium in tetrahydrofuran [36]. But when this reaction was run in the presence of a catalytic amount (1–3 mol%) of the aromatic electron-transfer agent 4,4′-di-*tert*-butylbiphenyl (DBB), the desired tertiary alcohol was obtained in 67% yield:

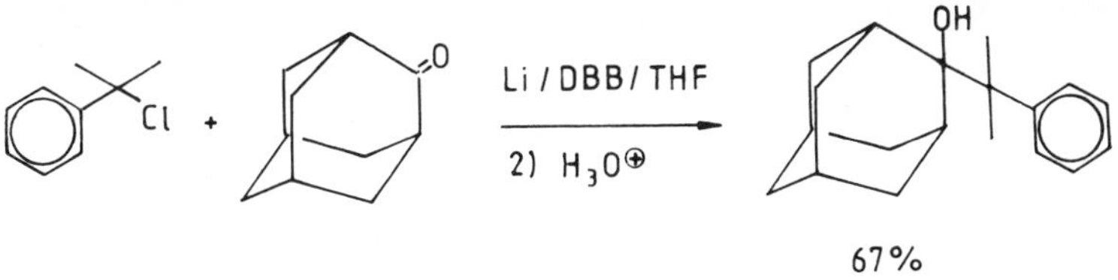

Li / DBB / THF

2) $H_3O^{\oplus}$

67%

No such catalyst was required for the synthesis of a carbocyclic analogue of quercetin with the same type of sterically crowded benzylic halide (A) [37]:

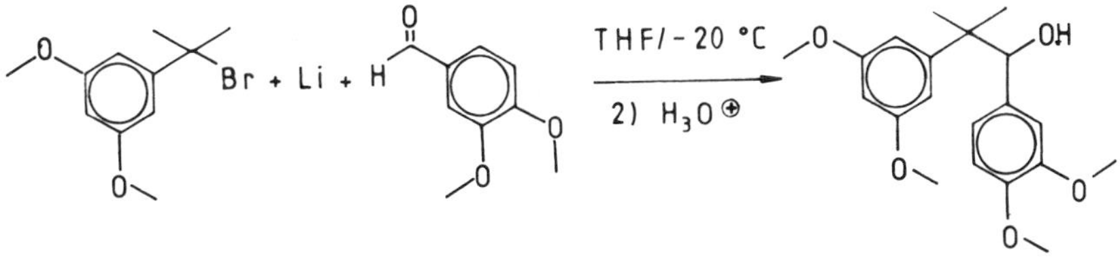

Br + Li + H

(A)

THF/ -20 °C

2) $H_3O^{\oplus}$

65%

Attempts to prepare the organolithium compound prior to coupling with the aldehyde, failed.

Other examples reported by the same authors of Barbier reactions with the bromide (A) and the corresponding trimethoxy chloride (A′) are:

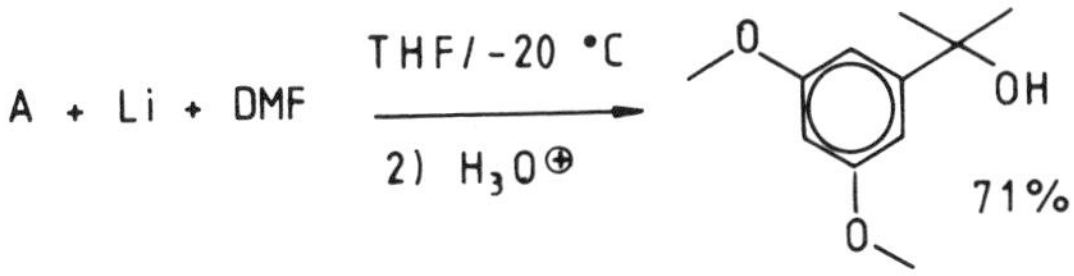

A + Li + DMF

THF/ -20 °C

2) $H_3O^{\oplus}$

71%

and

A′ + Li +

THF/ -20 °C

2) $H_3O^{\oplus}$

75%

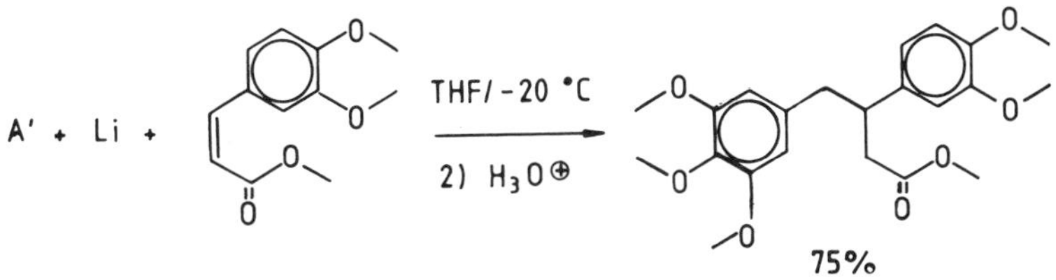

3.3.9 Li-Barbier Reactions with *n*-Butyllithium

A quite different version of the Li-Barbier reaction, which was recently published [38], has already been referred to on p. 86 *n*-Butyllithium was used to generate the intermediate lithium compound from *N*-(2-iodobenzyl) phenacylamine, achieving the first ever reported intramolecular Barbier reaction, with an aryl halide. As was the case with the iodonitrile (p. 86 of this Section), the metal-halogen exchange reaction of the *n*-butyllithium proved to be much faster than the carbonyl addition reaction.

n-BuLi

2) $H_3O^{\oplus}$

OH

69%

Eleven different iodides were cyclized in this manner with yields which varied from 50 to 86%.

Only in a few instances was the reduced product, resulting from hydrolysis of uncyclized aromatic lithium compound, obtained in relatively high yield (21%).

Reactions with bromides were less successful.

3.4 Zinc in Barbier-Type Reactions

3.4.1 Introduction

Saytzeff, Reformatsky and Simmons-Smith Reactions. The metal zinc played an important role in organic synthesis at the end of the nineteenth century, after Frankland [39, 40] had opened up this field of organometallic chemistry and after Saytzeff and coworkers [41, 42] had introduced a considerable number of syntheses with the aid of zinc in that same period (see Sect. 1.2).

After Barbier [6] introduced magnesium in the 'Saytzeff reaction', zinc lost its predominant position in this field of synthesis, although it remained the preferred metal in the Reformatsky reaction [43] (see Sect. 1.2.5).

However, in 1958, yet another powerful synthetic tool in which zinc plays a major role, was introduced to organic chemists i.e. the Simmons-Smith reaction [44, 45]. In this procedure, use is made of a carbenoid intermediate, formed by interaction of the metal (most commonly a zinc/copper alloy is used for this reaction) with diiodomethane:

$$CH_2I_2 + Zn/Cu \longrightarrow ICH_2ZnI$$

This reagent adds readily to carbon-carbon double bonds forming cyclopropane derivatives. This reaction is also often executed in a one-step fashion.

$$ICH_2ZnI + \text{cyclohexene} \longrightarrow \text{norcarane}$$

For reviews on the Simmons-Smith reaction see [46–48].

New Developments. A combination of new developments have recently stimulated a remarkable reintroduction of zinc as a coupling agent in traditional Barbier-type synthetic procedures.

As was made clear in Sect. 2.2 the greater reactivity of the (intermediate) Mg-Reformatsky reagent often causes interaction with substituted organic reactants leading to unplanned, and undesired side-reactions.

Hence, the use of the less reactive zinc metal in this widely used type of reaction is preferred.

The same holds for other synthetic paths in which metals are used. The lower reactivity of organozin intermediates makes them more selective reagents for substrates with reactive substituents.

One of the practical *dis*advantages of the metal zinc as a synthetic tool is its slow reactivity towards organo halides, which often causes unwanted side-reactions.

As mentioned in Sect. 1.2.2, in 1861 Pébal [49] used sulfuric acid to 'activate' zinc for his reactions and Rieth and Beilstein [50], one year later, used a 25% Zn/Na alloy for the activation of the metal. More recent is the use of Zn/Cu mixtures in reactions of metallic zinc [51, 52], among which is the above-mentioned Simmons-Smith reaction.

In the last three decades, three major developments occured for the preparation and handling of metals in heterogeneous reactions which have given zinc reagents (among other metals) new prospects in organic synthesis.

I A new method was introduced for the preparation of highly reactive, finely dispersed metals by reduction of their halides with the aid of alkali metals. These metals are named 'Rieke metals' (such as Rieke magnesium, Mg*, Rieke zinc, Zn*, etc.) after the inventor of the process [53–56]. Their reactivity is extremely high and reactions have been made possible which were simply unthinkable before.

II In that same period another type of highly reactive metal was introduced in synthetic organic chemistry: graphite intercalation compounds Mg/C_8, prepared by reduction of a metal halide by potassium graphitide, K/C_8 [57].

III The third powerful new tool in organic synthesis (see also p. 90 of this chapter) came with the introduction of sonochemistry, in which ultrasound waves are irradiated in reaction mixtures.
In 1980 sonication was first successfully introduced to synthesis through organometallic intermediates in Li-Barbier reactions [29] (see Sect. 3.3). This was rapidly followed by its use in 'one-pot' trifluoromethylation with zinc, in 1981 [58a], and in Reformatsky reactions in 1982 [58b].

In the field of modern organozinc chemistry, a surprisingly large number of Zn-Barbier reactions are published.

Reviews on the use of this and of several other metals in organic synthesis, such as tin and aluminium, can barely keep pace with these new developments.

3.4.2 'Saytzeff Reactions' with Propargylic Halides

Propargyl bromide reacts with (iodine-activated) zinc in the presence of (excess) cyclohexanone to yield 25–50% of the expected carbinol [59]:

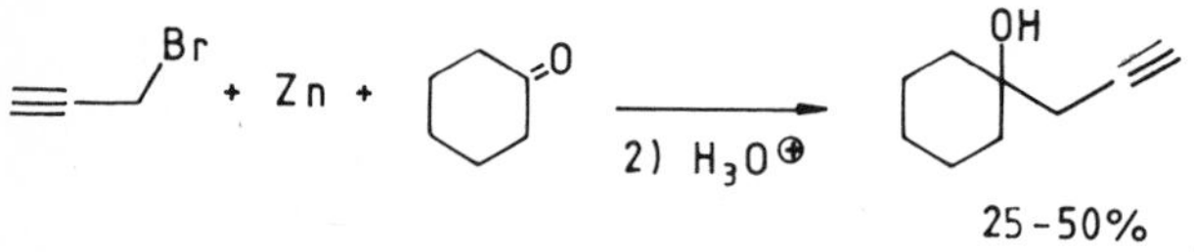

Br + Zn + =O → 2) $H_3O^{\oplus}$ OH 25-50%

When magnesium was used instead of zinc in this reaction 2-cyclohexylidenecyclohexanone was the only product isolated:

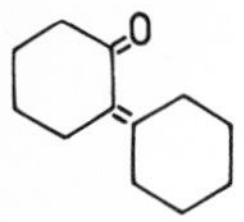

O

Further research in this field [60–62] showed that propargyl bromide undergoes this *Reformatsky type of reaction* with a variety of carbonyl compounds to give β,γ-unsaturated acetylenic carbinols in good yields. The remark was made that the *ethynyl group evidently activates the α-bromine atom in a similar manner to the ester group in the normal Reformatsky reaction.*

Among the aldehydes studied were butanal, 2-butenal and benzaldehyde (yields *ca.* 70%); among the ketones were cyclopentanone and -hexanone, acetophenone, benzophenone and methylheptenone [60].

Also substituted propargylic bromides such as (**1**) and (**2**) were used in these one-step reactions [60]. With benzaldehyde, yields of the carbinols were 90% and 70% respectively.

1 and 2

Saytzeff reactions of the vinylogous compounds (**3**) and (**4**)

3 and 4

with benzaldehyde or cyclohexanone gave rearranged carbinols in *ca.* 40% yields [61]:

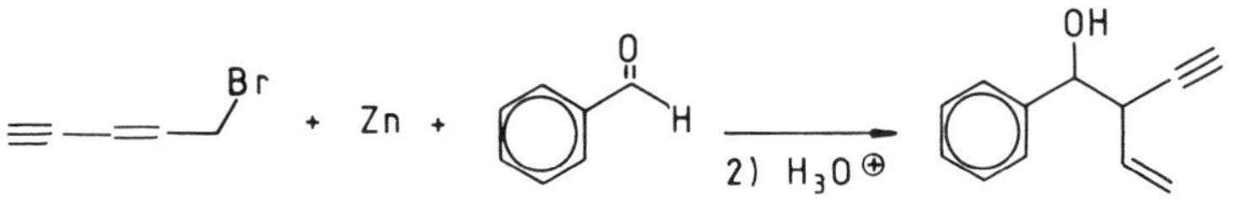

+ Zn + → 2) $H_3O^{\oplus}$

ca. 40%

This *novel reaction* [63] was also utilized to synthesize thiophene ethynilic carbinols in high yields:

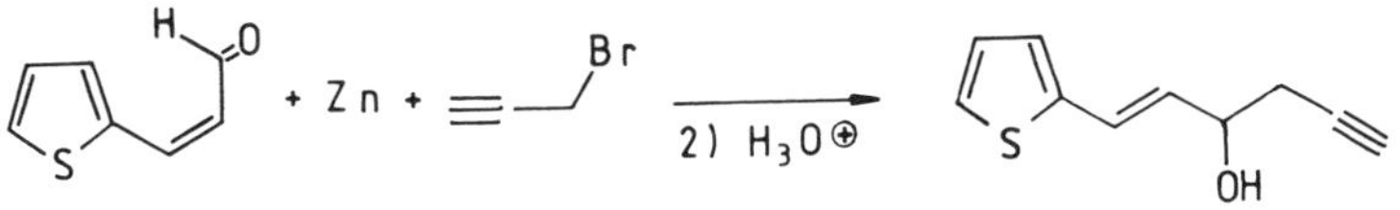

+ Zn + → 2) $H_3O^{\oplus}$

Systematic studies [64] of the products resulting from reduction – either by Zn/Cu in alcohol, by $LiAlH_4$ or by reaction with Mg followed by subsequent quenching of the reaction mixture with water – of substituted propargylic bromides, revealed that in each reaction the same mixture of compounds was obtained:

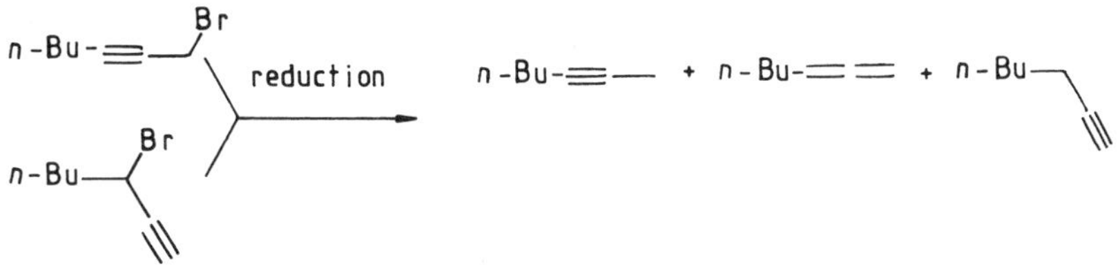

n-Bu-≡–CH₂Br / *n*-Bu–CH(Br)–C≡CH —reduction→ *n*-Bu-≡— + *n*-Bu-=C= + *n*-Bu—CH₂–C≡CH

These results shed a new light on earlier reports (by others) where the products were invariably identified as β-hydroxy-acetylenes.

The absence of allenic products can be explained by assuming insufficient analysis, or that the equilibrium in the intermediate organozinc compound favoured the formation of acetylenic products in their reaction with aldehydes or ketones.

On the basis of these new findings, it was advised to

carefully scrutinise the reaction products of propargylic halides.

No mention was made of the formation of allenic products in either of the following reports on 'Saytzeff reactions':

1. propargyl bromide with α-irone [65]):

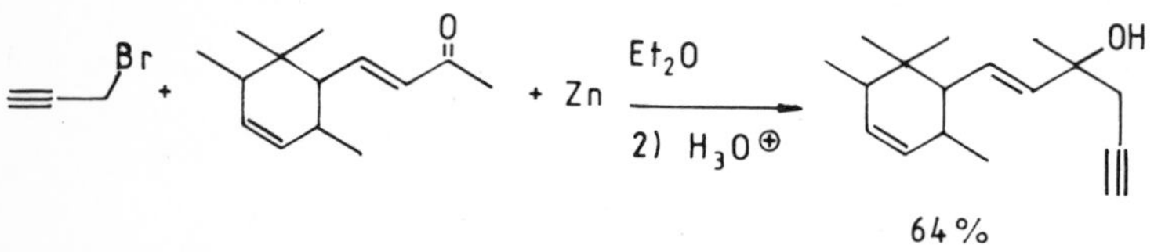

Br + O + Zn $\xrightarrow[\text{2) } H_3O^{\oplus}]{Et_2O}$ OH

64%

2. propargyl bromide with citronellideneacetone [66]):

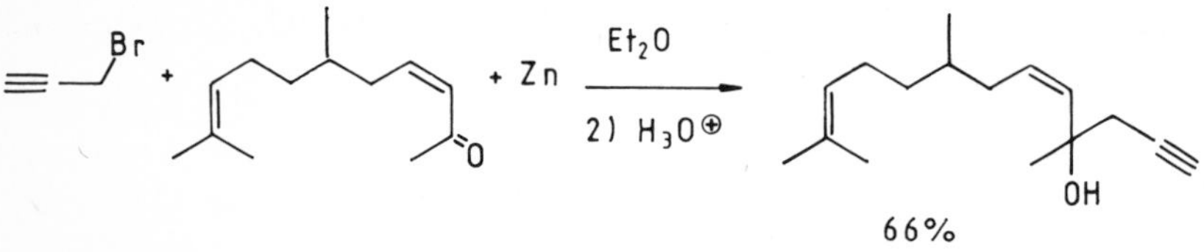

Br + O + Zn $\xrightarrow[\text{2) } H_3O^{\oplus}]{Et_2O}$ OH

66%

The reaction of 1-bromo-2-heptyne with zinc in the presence of 2,2-dimethylpropanal [67], however, did yield the allenic derivative as the main product together with small amounts of the acetylenic compound:

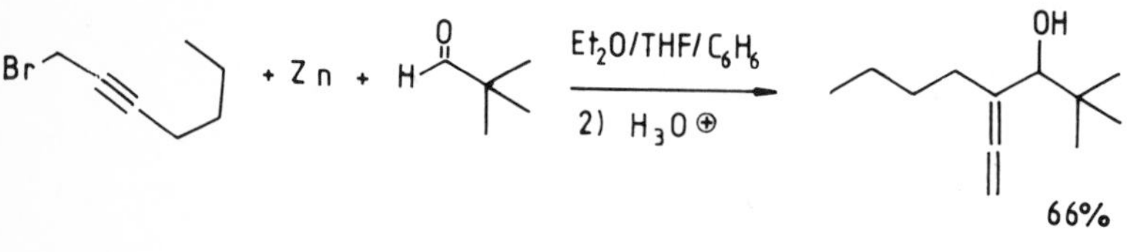

Br + Zn + H–C(O) $\xrightarrow[\text{2) } H_3O^{\oplus}]{Et_2O/THF/C_6H_6}$ OH

66%

The same was true for the reaction of 1-bromo-9-chloro-2-nonyne with heptanal [68]:

A few years later [69] the reaction of propargyl bromide with zinc in the presence of ethyl glyoxalate was reported to yield the propargylic derivative exclusively:

44%

As was recently found, the reaction of 1,1-difluoro-propargylic bromides [70] with zinc in THF yield the Wurtz-coupling product exclusively:

$$2\ R{-}{\equiv}{-}CF_2{-}Br + Zn \longrightarrow (R{-}{\equiv}{-}CF_2{-})_2$$

However, when aldehydes were present in the reaction mixture, the 'normal' Barbier-type reaction products were obtained (see Table 3.9):

3.4.3 Zn-Barbier Reactions with Perfluoro Haloalkanes

Perfluoroalkane carboxylic and sulfonic acids and their derivatives have been synthesized by Zn-Barbier reactions using iodoperfluoroalkanes in dimethylsulfoxide (DMSO) as solvent [71]. A Zn/Cu (100:1) couple was used for most of the reactions but other couples such as Zn-Pb, Zn-Cd or Zn-Hg led to the same results:

$$R_F{-}I + Zn + CO_2 \xrightarrow[2)\ H_3O^{\oplus}]{DMSO} R_FCO_2H$$

$$R_F{-}I + Zn + SO_2 \xrightarrow[2)\ Cl_2,\ CH_3OH]{DMSO} R_FSO_2Cl$$

Yields varied from 40% to 80%.

Table 3.9. Results of Barbier Type Reactions with the Presence of Aldehydes in the Reaction Mixture [70]

$$R{-}C{\equiv}C{-}CF_2{-}Br + Zn + R'CHO \xrightarrow[2)\ H_3O^{\oplus}]{THF} R{-}C{\equiv}C{-}CF_2{-}CH(OH)R'$$

R-	R'CHO	Yield (%)
C_6H_5	C_6H_5CHO	78
"	C_6H_5–(CH₂)ₙ–CHO	82
"	2,2-dimethyl-1,3-dioxolane–CHO	78
n-C_5H_{11}-	C_6H_5CHO	38
"	2,2-dimethyl-1,3-dioxolane–CHO	50
"	$THP{-}O{-}CH_2CH{=}CHCHO$	35
$(C_2H_5)_3Si$-	$C_6H_5CH_2CH_2CHO$	47

It is worth noting that e.g. perfluoropropylzinc iodide, C_3F_7ZnI, is unre active towards carbon dioxide under normal conditions [72]. The assumption was therefore made

that for these one-step procedures the reactivity of the organometallic intermediate is increased on the metallic surface and that the solvent and the couple play an important role in allowing the reaction to occur.

Recent work demonstrated that a more efficient and cheaper method to synthesize trifluoromethanesulfonic acid (triflic acid) is to use bromo instead of iodo trifluoromethane [72a] under a slight overpressure of *ca.* 3 atmosphere with DMF as the solvent.

Numerous metals were tested in this reaction: zinc, aluminium, cadmium and manganese were effective, whereas magnesium, tin, iron, nickel and cobalt failed to bring about condensation:

$$CF_3\text{-}Br + Zn(Al,\ Cd\ or\ Mn) + SO_2 \xrightarrow[2)\ H_3O^{\oplus}]{DMF} CF_3SO_2H$$

As mentioned in the Introduction to this Section, one of the important new techniques which has recently been applied in organometallic syntheses is sonication, the irradiation with ultrasonic waves.

Soon after its first application [29] (see p. 90 in this Section) it was found that the trifluoromethylation of carbonyl compounds was readily accomplished by the *a one-pot* reaction of iodotrifluoromethane, zinc and a carbonyl compound in DMF or THF at room temperature [58b].

Typical results are presented in Table 3.10.

The versatility of this reaction was further demonstrated in a study [73] of the reaction of iodo- and bromoperfluoroalkanes with zinc under sonication in the presence of:

a) aldehydes and ketones and a titanium catalyst,
b) vinylic and allylic halides (with a palladium catalyst) to give the Wurtz-coupling product,
c) enamines which yield ketones (also with a titanium catalyst).

Some typical results are listed in Tables 3.11, 3.12, 3.13 and 3.14.

Even without sonication Zn-Barbier reactions with fluorinated organo halides proved to be very effective. 1, 1, 1-Trifluoro-2, 2, 2-trichloroethane reacted with zinc in the presence of aldehydes in DMF [74–76].

Different types of products were obtained when an excess of the metal was used together with either acetic acid anhydride or aluminium chloride in catalytic amounts:

$$RCHO + Cl_3C\text{-}CF_3 \xrightarrow{DMF} \begin{cases} \xrightarrow{Zn\ (1\ mol)} R\text{-}CH(OH)\text{-}CCl_2CF_3 \\ \xrightarrow{Zn(>2\ mol)/Ac_2O} R\text{-}CH{=}C(Cl)\text{-}CF_3 \\ \xrightarrow{Zn(>2\ mol)/cat.\ AlCl_3} R\text{-}CH(OH)\text{-}C(Cl){=}CF_2 \end{cases}$$

Table 3.10. Trifluoromethylation of carbonyl compounds [58b]

$$CF_3\text{-}I + Zn + \text{Carbonyl Compound} \xrightarrow[2)\ H_3O^{\oplus}]{DMF/)))} \text{Product}$$

Carbonyl Compound	Product	Yield (%)
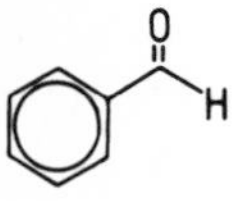	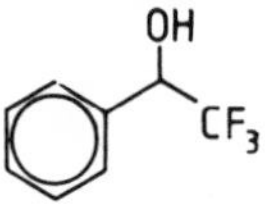	72
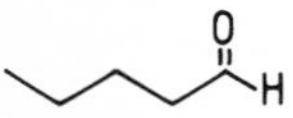	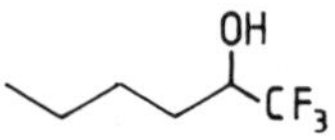	61
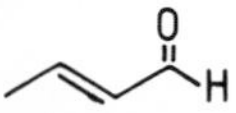	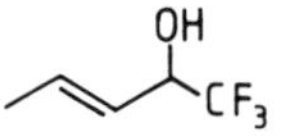	62
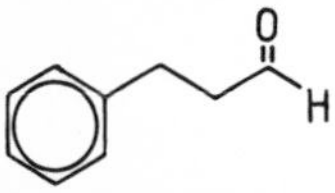	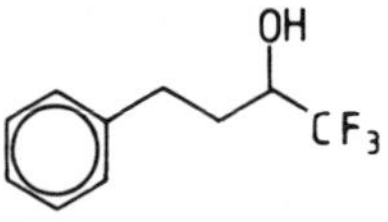	68
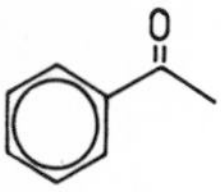	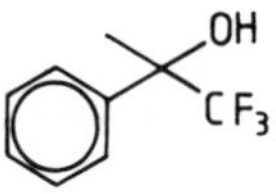	55
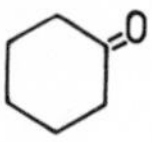	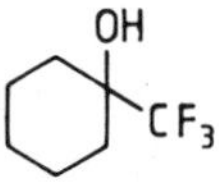	48
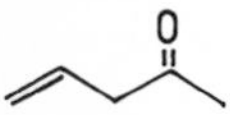	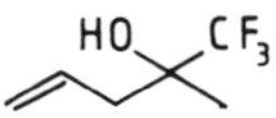	45

The following aldehydes were applied in these reactions:

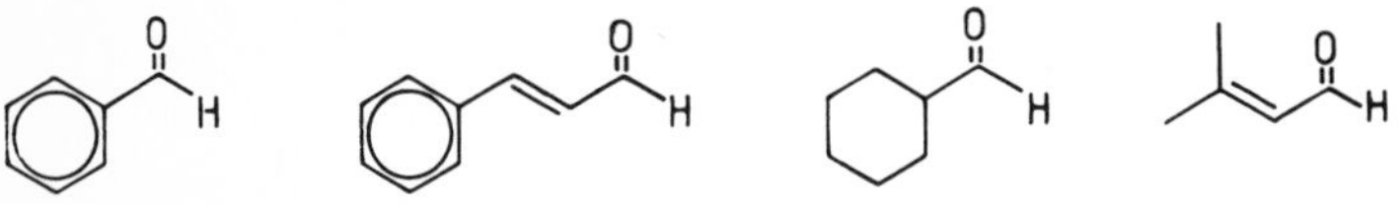

Yields varied from *ca* 50% to well over 80%.

Table 3.11. Perfluoroalkylation of benzaldehyde [73]

$$R_F\text{-}X + Zn + PhCHO \xrightarrow[2)\ H_3O^{\oplus}]{DMF / \cdot))} PhCH(OH)R_F$$

R_F-X	Yield of the alcohol (%)
CF_3-I	72
CF_3-Br	56
C_2F_5-I	67
n-C_3F_7-I	58
i-C_3F_7-I	56
n-C_4F_9-I	61
n-C_4F_9-Br	49
n-C_6F_{13}-I	66
n-C_8F_{17}-I	58
n-C_8F_{17}-Br	32

The yields of alcohols could be enhanced in several instances when sonication was applied; the same was true with various catalysts such as CuCl or $PdCl_2$- and $NiCl_2$-(triphenylphosphine)$_2$.

The use of Pd(II)- or Ni(II)-triphenylphosphine catalysts for Zn-Barbier reactions in DMF had been applied earlier for the reaction of iodoperfluoroalkanes with several different types of aldehydes [77]:

$$R_F\text{-}I + Zn + HC(O)R \xrightarrow[2)\ H_3O^{\oplus}]{Ni\text{- or } Pd\text{-catal.}/DMF} R_FCH(OH)R \quad 20-60\%$$

A radical mechanism was proposed for this reaction in which zinc was suggested to first produce Pd^0 or Ni^0. $R_F\cdot$ radicals could then either react with

Table 3.12. Perfluoroalkylation of carbonyl compounds with Zn in the presence of a catalyst [73]

$$\mathrm{RCHO} + \mathrm{Zn} + \mathrm{R_FX} + \text{catalyst} \xrightarrow[\text{2) } H_3O^{\oplus}]{\text{DMF/ ...)}} \mathrm{R{-}CH(R_F)OH}$$

R	R_F-X	Catalyst	Yield of the alcohol (%)
phenyl	CF_3-I	-	13
"	"	Cp_2TiCl_2	36
"	CF_3-Br	"	33
"	n-C_4F_9-I	-	12
"	"	Cp_2TiCl_2	38
vinyl	CF_3-I	-	18
"	"	Cp_2TiCl_2	41
"	CF_3Br	"	36

the aldehyde to produce the desired product, with each other in a Wurtz-coupling reaction, or with ZnI^+ to give the perfluoroalkylzinc iodide.

The new reaction could also be executed with the much cheaper bromotrifluoromethane instead of the iodo compound and resulted in about the same yield.

Trifluoromethyl alcohols have been synthesized in pyridine via Zn-Barbier reactions of carbonyl compounds with trifluoromethyl halides under a slight pressure [78, 79]. Some results are listed in Table 3.15

3.4.4 Other Zn-Barbier Reactions; Solvents, Activating Agents

It will be demonstrated in this Section that the application of unusual solvents and/or activating agents has opened up new synthetic possibilities through Zn-Barbier reaction.

Table 3.13. Perfluoroalkylation of Vinylic Halides with tetrakis (triphenylphosphine) palladium as a catalyst [73]

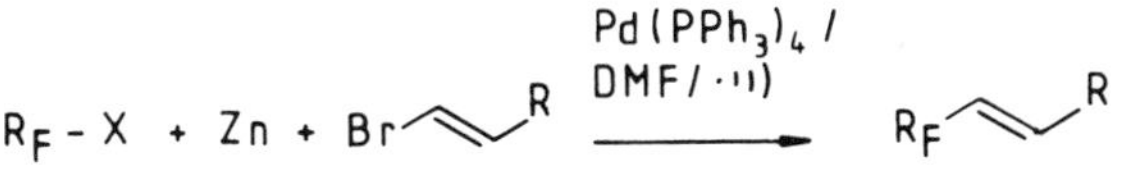

$$R_F-X + Zn + Br\text{-}CH{=}CH\text{-}R \xrightarrow[\text{DMF/))) }]{Pd(PPh_3)_4\,/} R_F\text{-}CH{=}CH\text{-}R$$

R_F-X	R-	Yield (%)
CF_3-I	 phenyl	65
CF_3-Br	"	53
C_2F_5-I	"	47
n-C_3F_7-I	"	66
i-C_3F_7-I	"	72
n-C_4F_9-I	"	62
n-C_4F_9-Br	"	32
CF_3-I	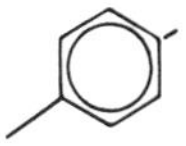 methylphenyl	67
CF_3-Br	"	41
n-C_3F_7-I	"	68

Trimethylchlorosilane. The use of trimethylchlorosilane (TMSCl) as an activator for reactions of metallic zinc (5 molar equivalents of TMSCl and 10 molar equivalents of Zn in diethyl ether) was recently introduced [80] in the following reaction:

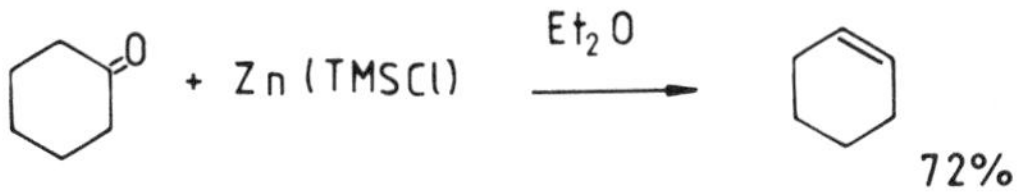

$$\text{cyclohexanone} + Zn\,(TMSCl) \xrightarrow{Et_2O} \text{cyclohexene} \quad 72\%$$

Table 3.14. Palladium-catalyzed perfluoroalkaylation of allylic halides [73]

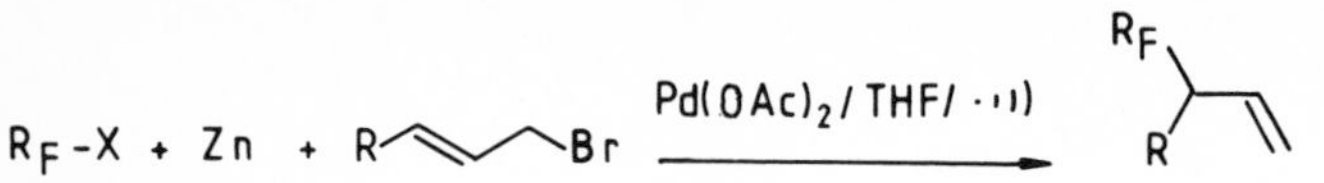

R_F-X + Zn + R–CH=CH–CH$_2$Br $\xrightarrow{Pd(OAc)_2/THF/\cdot))}$ R(R$_F$)CH–CH=CH$_2$

R_F-X	R-	Yield (%)
CF_3-I	 C_6H_5-	51
CF_3-Br	"	36
C_2F_5-I	"	42
n-C_3F_7-I	"	71
i-C_3F_7-I	"	78
n-C_4F_9-I	CH_3-	68
n-C_4F_9-Br	"	37
n-C_6F_{13}-I	"	56

Several steroid ketones were reported to be deoxygenated in good yields by this method [80a]: e.g. 5α-cholestan-3-one was converted in 69% yield to 5α-cholest-2-ene:

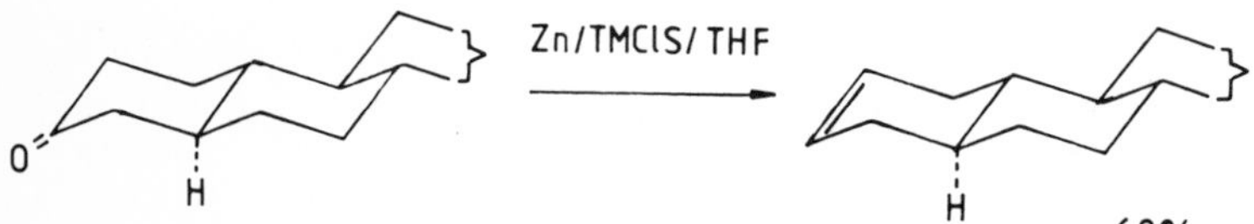

Zn/TMClS/THF

69%

Zn/TMSCl was also used for a new method of five-membered ring annelation [81]. The mechanism probably involves free radical generation from ketones, followed by internal addition to a variety of π-unsaturated functions X

Table 3.15. Trifluoromethylation of Carbonyl Compounds [78], [79]

Carbonyl Compound + Zn + CF_3-Br $\xrightarrow[\text{2) } H_3O^{\oplus}]{\text{2-3 atm./pyr.}}$ Product

Carbonyl Compound	Product	Yield (%)
(benzaldehyde)	(PhCH(OH)CF_3)	52
(cyclohexanone)	(1-CF_3-cyclohexanol)	20
(ethyl pyruvate)	(HO, CF_3 adduct)	35
(phthalic anhydride)	(HO, CF_3 lactone)	61

such as $=CH_2$, $\equiv CH$, $=O$, $\equiv N$:

$\xrightarrow[\text{TMSCl}]{\text{Zn}}$ OTMS ... X $\longrightarrow$ OTMS ... X• $\xrightarrow{\text{RH}}$ OTMS ... XH

After the successful application of Zn/TMSCl in the Reformatsky reaction [82] (see also more recent work in this field [83–85]), the same combination was applied in Zn-Barbier reactions with allylic bromides in diethyl ether [86]:

Br + (ketone) $\xrightarrow[\text{2) } H_3O^{\oplus}]{\text{Zn/TMSCl}}$ OH 72%

Table 3.16 presents results of such reactions with different substrates.

Table 3.16. Results of Barbier-type reactions with Zn/TMSCl [86]

Bromide	Carbonyl compound	Product(s)	Yield %
Br	O, H	OH	50
"	O, H	OH	55
"	O, H	OH	70
"	O	OH	79
"	O	HO	69
"	O	HO	72
"	O, O	OH	75
"	O, H, O	OH; O, H, O (77 : 23)	-
Br	O	OH	77
Br	O, H	OH	43
Br, n-C_4H_9	O	n-C_4H_9, OH; n-C_4H_9, OH (82 : 18)	53

From the following reaction with cyclohexene, it was concluded that the deoxygenation of aldehydes and ketones by Zn/Cu and TMSCl in diethyl ether takes place via a carbenoid intermediate [86a].

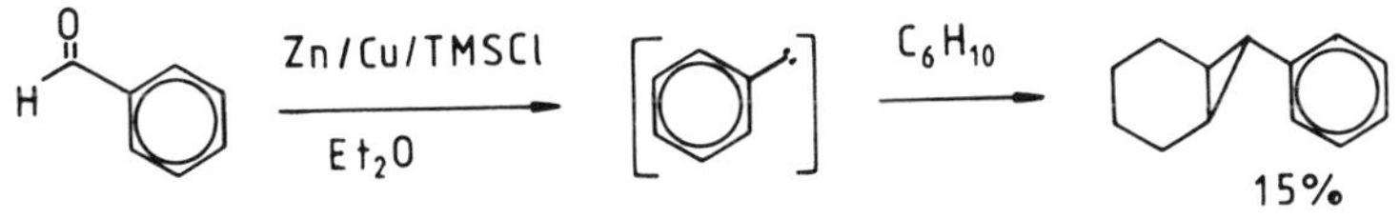

Zn/Cu/TMSCl
Et_2O
C_6H_{10}
15%

Using dimethyldichlorosilane (DMDClSi) much higher yields of carbenoid intermediates are observed:

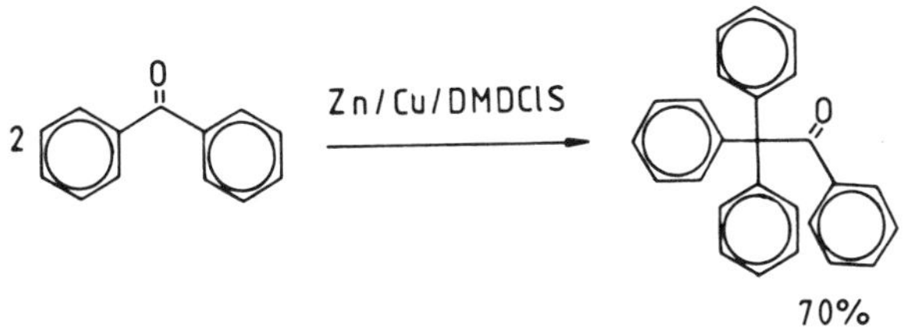

2
Zn/Cu/DMDClS
70%

Alcohols as the Solvent. Rather surprising was the introduction of alkanols such as ethanol or *tert*-butyl alcohol as the solvent in Zn-Barbier reactions with allylic halides [87]:

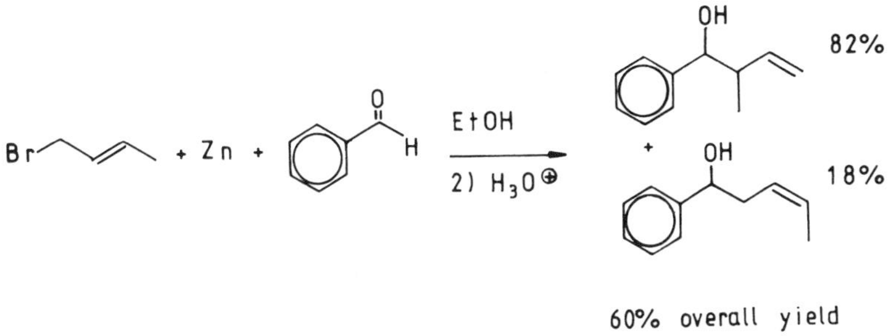

Br + Zn +
EtOH
2) $H_3O^{\oplus}$
OH
82%
+
OH
18%
60% overall yield

The yield decreases as the reactivity of the carbonyl compound decreases but can be improved by employing two equivalents of the allyl component and zinc. Some results with allyl bromide are presented in Table 3.17.

To insure that allylzinc bromide was actually an intermediate in these reactions a sample was prepared in THF.

Table 3.17. Some results of Zn-Barbier reactions of allylbromide with carbonyl compounds in alcoholic solution [87]

Carbonyl Compound	Yields (%) obtained in	
	EtOH	t-BuOH
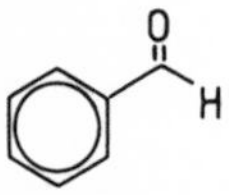	66	71
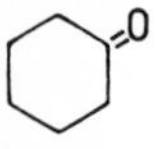	42	62
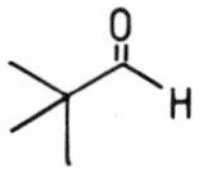	39	33
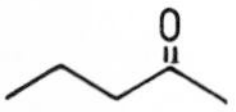	23	61
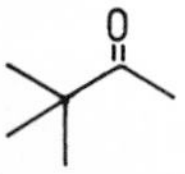	trace	22

Addition of two different portions of this solution

a) to benzaldehyde, dissolved in ethanol, and
b) to benzaldehyde, dissolved in THF, gave the expected product (1-phenyl-3-buten-1-ol) in 60 and 80% yield respectively.

Water as the (Co-)Solvent. In view of the reactivity of organozinc compounds towards 'acidic' hydrogens, the use of water as the (co-)solvent in Zn-Barbier reactions is even more surprising than the use of alcohols.

There has been a gradual approach to the use of water (a very attractive solvent in view of its low cost, environmental safety and low inflammability) as a (co-)solvent in organometallic reactions.

After tribenzyltin chloride was prepared by reaction of benzyl chloride with tin in boiling water [88], Sn-Barbier reactions with allylic bromides (see also Sect. 3.5) were successfully carried out in a 1:1 mixture of water and diethyl ether with a catalytic amount of hydrobromic acid [89].

Water as the (co-)solvent in Zn-Barbier reactions followed soon after albeit though only allylic bromides could be used [90].

Allyl bromide and benzaldehyde gave 100% yield of the expected alcohol after 45 min in a 5:1 mixture of saturated aq. NH_4Cl/THF at room temperature (for the activation of zinc by NH_4Cl see [91]):

Br + Zn + PhCHO —aq. NH_4Cl/THF/R.T.; 2) $H_3O^{\oplus}$→ PhCH(OH)CH₂CH=CH₂ 100%

In aq. NH_4Cl without THF the yield of the same reaction was 61%. In aq. NaCl/THF no product at all could be isolated. The use of a 5:1 H_2O/THF mixture as the solvent for the same reaction yielded only 48% of the product when mechanical stirring was applied. Under sonication the yield of this last reaction was even lower: 23%.

Other results of interest are:

a) Relatively low yields (58% and 53%) were obtained with allylic chlorides and with ketones.
   On the other hand a 95% yield was reported for the following reaction under the same conditions:

Cl + Zn + H(C=O) —sat. NH_4Cl/THF; 2) $H_3O^{\oplus}$→ OH 95%

b) although the reaction of crotyl bromide with cyclohexanone yielded the expected product quantitatively the reaction of allyl bromide with 2-methylhept-2-en-6-one gave not more than 74% of the expected product.

A reasonable mechanism for these reactions in aqueous solutions could not be proposed but a classical organometallic one seems quite improbable.

Competition reactions between aldehydes and ketones [92, 93] demonstrated a strong preference for reactions with the aldehyde.

The positive results obtained with allylic halides could not be achieved with other halides such as benzyl and butyl bromides.

It should also be underlined that in the Zn-Barbier reaction with allylic halides the use of sonication did not necessarily lead to better results.

The following comments are of note:

a) no important differences were observed in results obtained under anhydrous conditions.

b) thus the reaction medium has no marked influence on the steric course of the reaction; further revealing that the medium can not be directly involved in the transition state.
The conclusion can be drawn that the reacting species is probably adsorbed on the metal surface. This view is supported by the complete absence of asymmetric induction observed in a reaction where a chiral inductor (mannitol), was present in the reaction mixture.

c) the presence of free hydroxyl groups in the substrate did not affect the general course of the reaction:

OH, H, O + Zn + Br — sat. NH_4Cl/THF; 2) $H_3O^{\oplus}$ → OH, OH 75%

In an interesting modification of the aqueous NH_4Cl/THF combination, a solid organic support was recently used instead of the cosolvent THF [94]. Reaction times under these conditions varied from 0.5 to 16 h.

The simplicity of the allylation reaction and its workup makes this procedure highly attractive since the organic phase (C-18 silica, reverse-phase chromatography support) can be reused and the solvent (water) is *uniquelyenvironmentally* safe!

Another great advantage of this type of reaction is that additional hydroxy functional groups could be present in the substrate:

2 Br + 2Zn + HO, O, H, H, O — sat. NH_4Cl/ C-18 silica; 2) $H_3O^{\oplus}$ → HO, OH, OH 60%

and

Br + Zn + O, OH — sat. NH_4Cl / C-18 silica; 2) $H_3O^{\oplus}$ → OH, OH 90%

Addition to Double Bonds in Zn-Barbier Reaction. Acetonitrile was used as the solvent in a surprising type of Zn-Barbier reaction, in which an alkyl iodide was reacted with (unactivated) zinc in the presence of both an 'activated' olefin and a carbonyl compound [95].

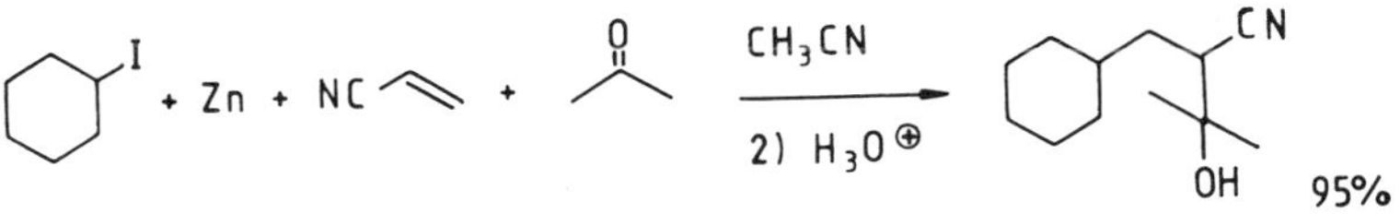

In this *zinc-promoted joining reaction* (as the authors named it) acetic acid anhydride behaved in a similar manner to ketones and aldehydes:

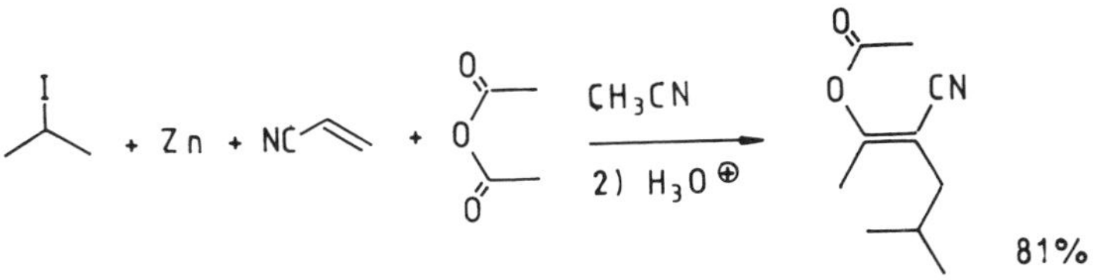

With a suitable iodoketone a cycloaddition reaction took place in reasonable yield:

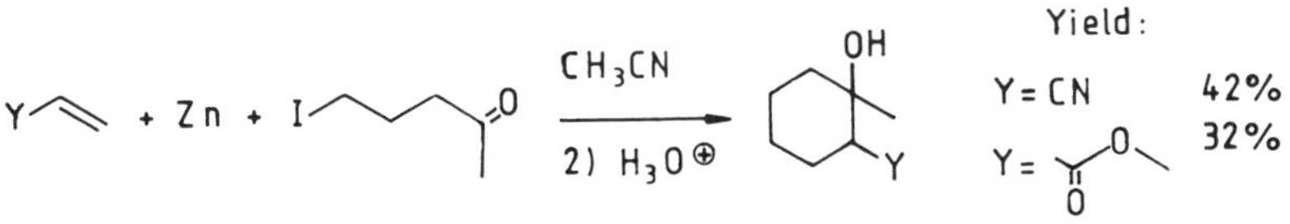

Addition reactions to the C=N double bond of isoquinolinium iodides were studied in one-step processes with benzylic halides and Zn in acetonitrile as the solvent [96].

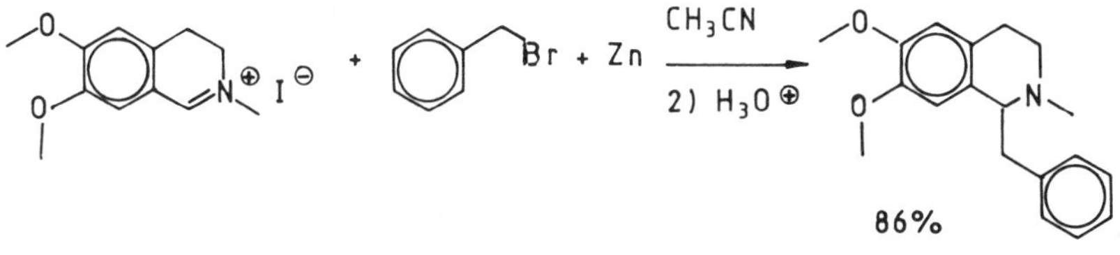

The same type of addition reaction was found with allylic halides and with 1- and 2-iodopropane.

In a series of studies [97–99] the one-step – Reformatsky type – addition reaction to carbon–carbon double and triple bonds was investigated with ultrasound irradiation in which zinc was used together with activated allylic bromides such as

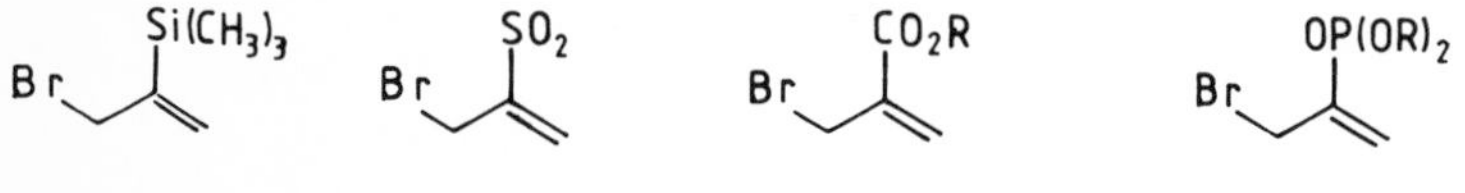

In the presence of vinyldiphenylphosphineoxide, haloalkanes react with a 1:1 zinc-copper couple (which was to be preferred over NH_4Cl-activated zinc) under sonication in a 9:1 ethanol/water mixture to give C=C double bond addition [100]:

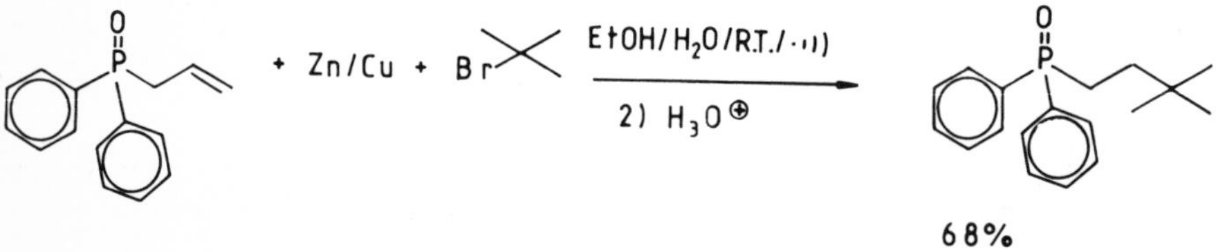

A conjugate addition of a hydroxy–substituted (!) iodide and 1-buten-3-one was accomplished in excellent yield using a 4:1 Zn/CuI mixture and sonication [100a]:

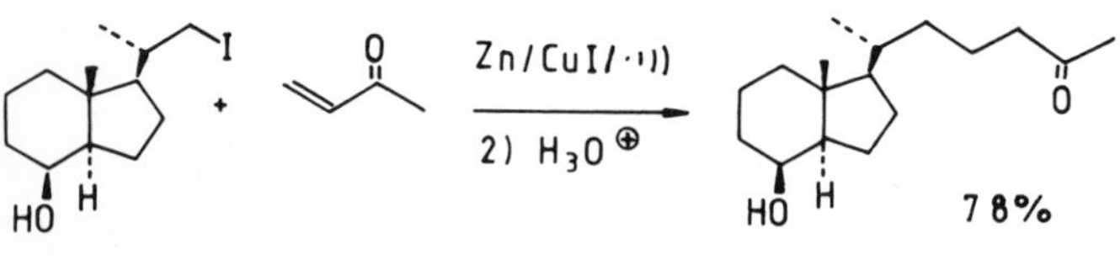

1-Bromoadamantane (1-AdBr) reacts with Zn/Cu under sonication in the presence of activated double bonds to give conjugate addition products with varying success [100b]

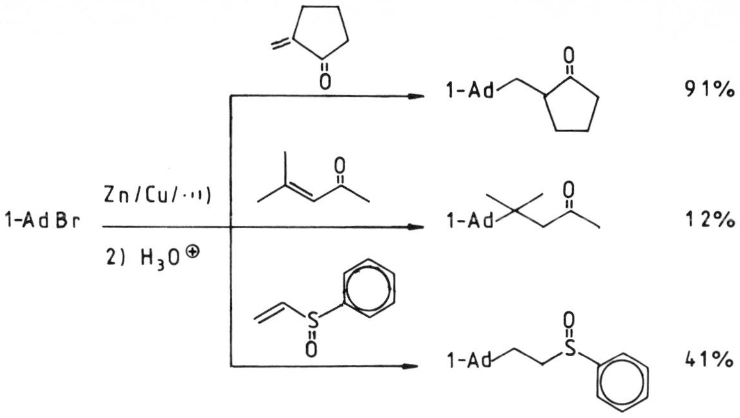

Substituted nortricyclenes were obtained in good yields when aryl or benzyl halides were reacted with norbornadiene in the presence of paladium complexes and zinc powder [101]. Surprisingly water was present in the reaction mixture, with THF as the major solvent. Some of the results reported were as presented in Table 3.18.

Table 3.18. Yields of substituted nortricyclenes from Pd-catalysed Zn-Barbier reactions [101]

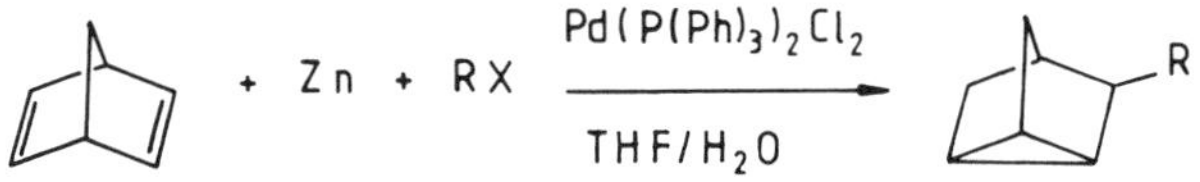

RX	Yield (%)	
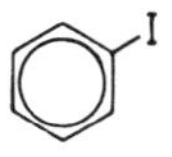	86	
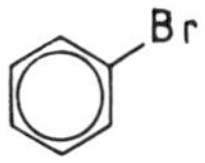	54	
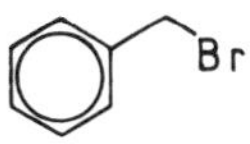	35	
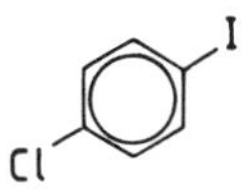	85	(the chloro substituent was left untouched)

Zn-Barbier Reactions with Nitriles. The reaction of allylic halides with zinc in the presence of nitriles, was first studied by Blaise [102] in 1901. Later investigations showed that much better yields of the β-unsaturated ketones could be obtained if benzene was employed as the solvent and when a zinc-silver couple was used [103, 104] (prepared from zinc and 0.1% silver acetate in methanoic acid).

The results of the following reaction are presented in Table 3.19:

Table 3.19. Results from Barbier-type reactions of allylic bromides and organic nitriles with Zn/Ag [104]

Nitrile	Bromide	Solvent	Product	Yield(%)
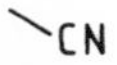 CN	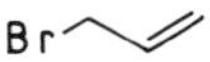 Br	ether/THF (90 : 10)	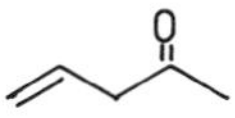 O	75
"	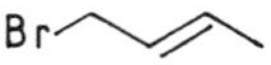 Br	benzene	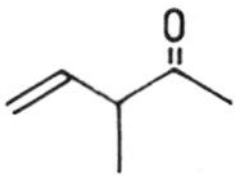 O	78
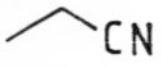 CN	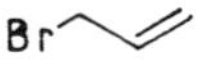 Br	ether	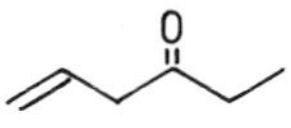 O	80
"	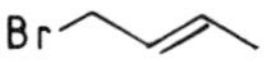 Br	benzene	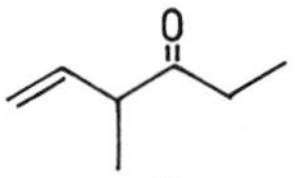 O	80
"	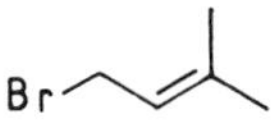 Br	THF	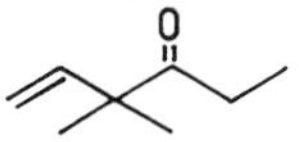 O	71
 CN	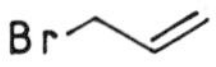 Br	ether	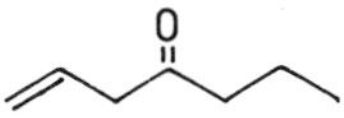 O	80
"	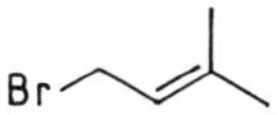 Br	THF	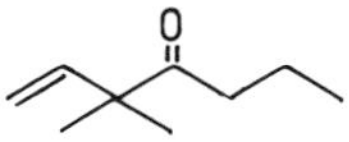 O	68
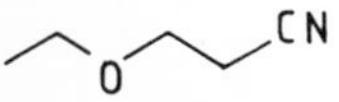 O CN	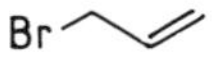 Br	ether	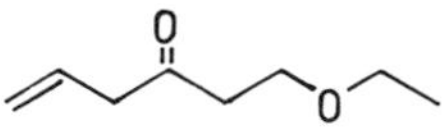 O O	78
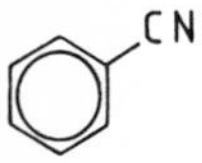 CN	"	benzene	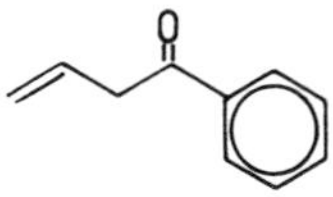 O	0
"	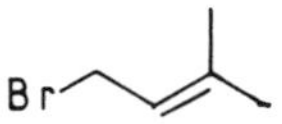 Br	THF	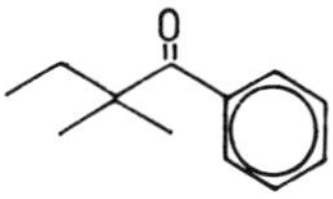 O	67

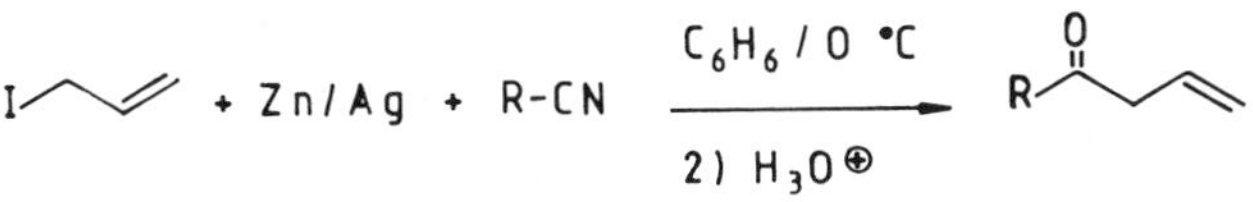

In an earlier investigation (1957, [105]) the Zn-Barbier reaction of cyano-ethane and allyl bromide with unactivated zinc in benzene gave only a 20% yield of the expected product.

Replacement of zinc by magnesium and the use of diethyl ether as the solvent, led to complete failure; only resinous material was obtained.

Several more complicated allylic bromides were used for this Zn-Barbier reaction in a later publication [106], leading to the formation of terpene–related dienones:

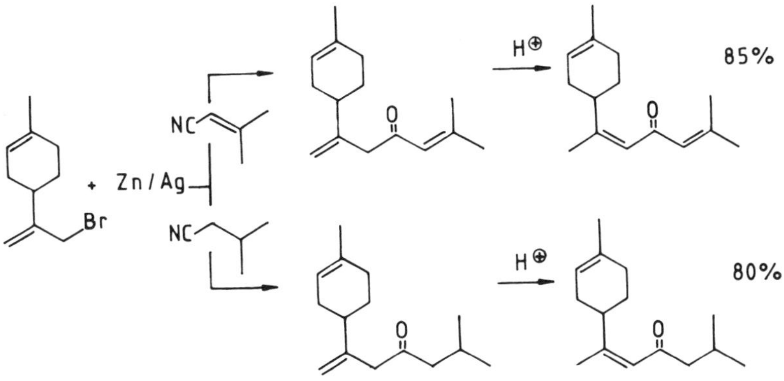

Zn-Barbier Reactions with Di-Bromides. Reaction of *ortho*-substituted bis(bromomethyl) aromatic compounds with (NH_4Cl-activated) zinc in DMF probably leads to the formation of reactive *o*-xylylene intermediates that react with methyl vinyl ketone to form a conjugate addition reaction product [91]:

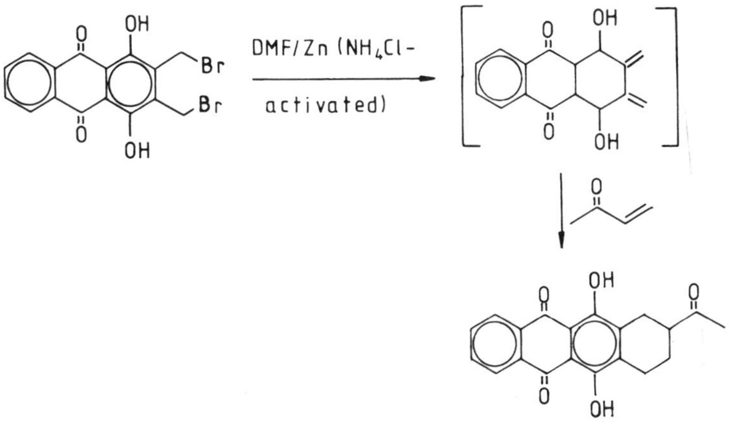

1,2-Bisbromomethylbenzene was reacted with NH_4Cl-activated zinc in the presence of a dienophile under sonication [107] to give high yields of the expected Diels-Alder products. Reaction temperatures were 20–25 °C and reaction times 12–15 h. The results are listed in Table 3.20.

Highly hindered bicyclo[3.2.1]oct-6-en-3-ones, including a variety of spiroannulated derivatives, have been synthesized through a Zn/Cu promoted reaction of α,α'-dibromo ketones with 1,3-dienes in dioxan under sonication [108].

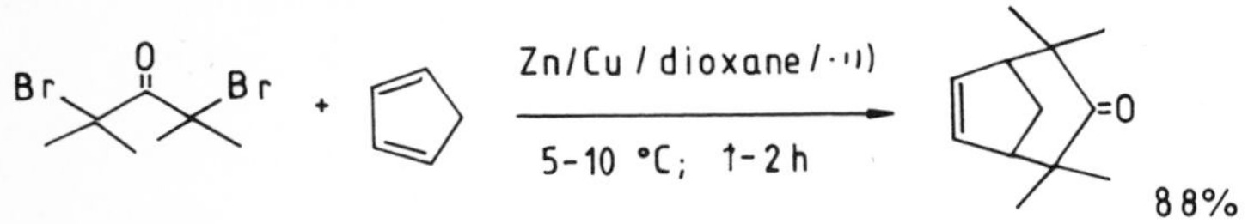

Zn/Cu / dioxane / ·))) ; 5–10 °C; 1–2 h ; 88%

With furan the yield was 91%. Cyclic ketones also gave high yields (α,α'-dibromodicyclopentyl ketone: 71%). Amongst various solvents examined, di-

Table 3.20. Results of Barbier-type reactions of 1,2-bisbromomethylbenzene with zinc and a dienophile [107]

1,2-bis(bromomethyl)benzene + Zn + dienophile —(dioxane, ·)))→ product

dienophile	product	yield (%)
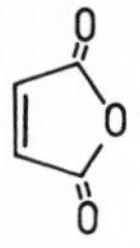 maleic anhydride	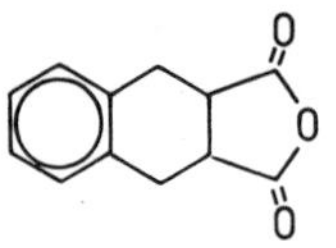	89
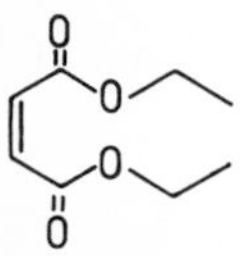 diethyl maleate	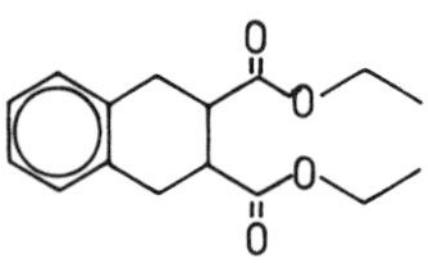	70
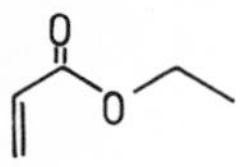 ethyl acrylate	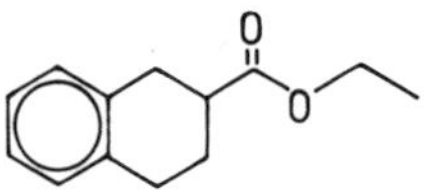	67
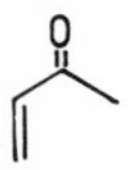 methyl vinyl ketone	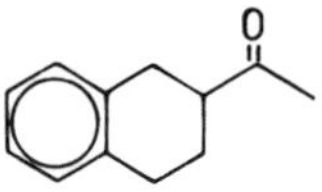	87

oxan was found to be the solvent of choice. The reaction gave poor yields in diethyl ether or benzene and proceeded erratically in THF.

Zinc-assisted condensation with lower valent titanium is demonstrated in the following reactions [109]:

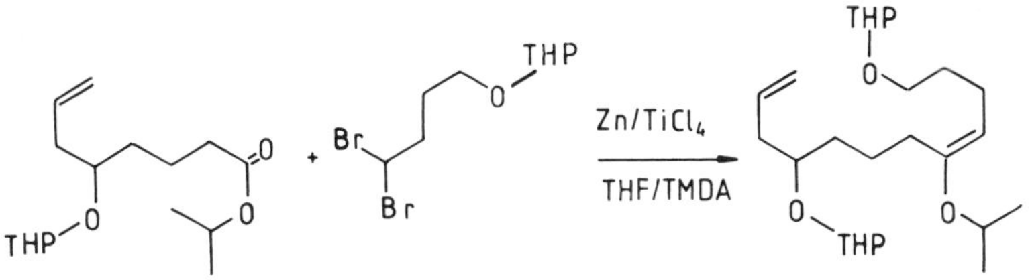

Zn/TiCl$_4$

THF/TMDA

and

Zn/TiCl$_4$

THF/TMDA

15%

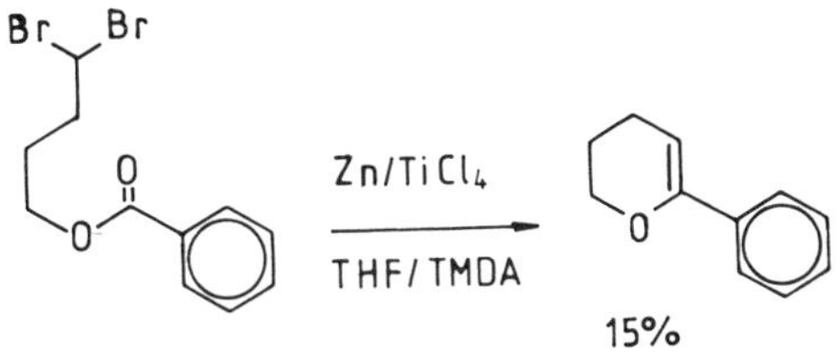

A review article [110] on condensation reactions of lower valent titanium reveals the wide use of zinc in one-step processes.

Et_2AlCl-Assisted Carbonyl Addition. A new method [111] for the preparation of β-hydroxy-ketones involves the coupled attack of dialkylaluminium chloride and zinc on an α-bromoketone. An aluminium enolate is generated which gives a facile addition to another carbonyl compound, present in the system (illustrated below):

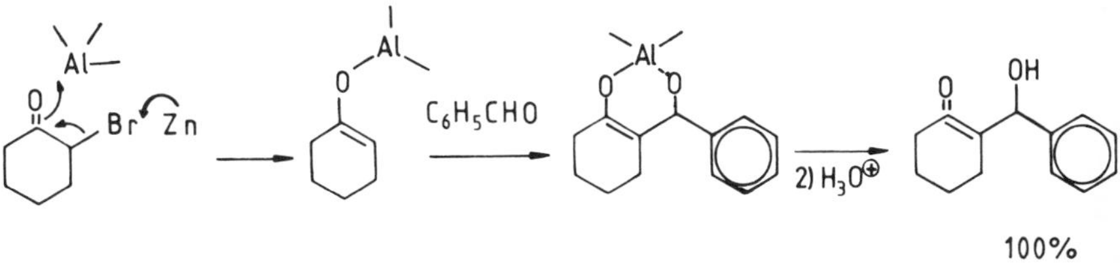

C_6H_5CHO

2) $H_3O^{\oplus}$

100%

Electroassisted Barbier Reactions. In Sect. 2.4 the so-called *electro-assisted Barbier reaction* was mentioned in which a wide range of alcohols were synthesized from organo halides, carbonyl compounds and a sacrificial magnesium anode.

Zinc can be used as the metal for such reactions [112].

A recent report [113] of this type of Barbier reaction mentions allylation of imines with allyl bromide in a $PbBr_2/Bu_4NBr$/THF-Zn(anode)-Pt(cathode) system, leading to the overall reaction shown below [113]:

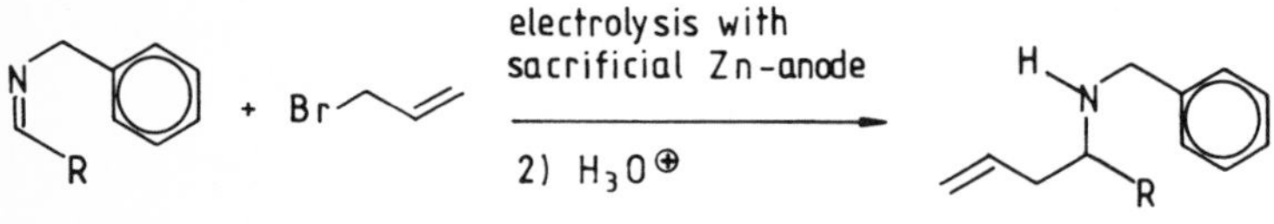

electrolysis with sacrificial Zn-anode

2) $H_3O^{\oplus}$

As in the previous report aluminium could also be used as the sacrificial electrode.

Activated Zinc in Barbier-Type Reactions. Throughout this Section the activation of zinc metal has been described in order to make Barbier-type reactions proceed in an improved fashion.

Zinc has been activated with iodine, silver, ammonium chloride, copper etc. This last paragraph deals with two special forms of activated zinc

a. finely divided zinc as introduced in the 1970s Rieke [53–56], and
b. zinc-graphite intercallation compounds, Zn^{o}/Gr [57].

Rieke Zinc in Barbier Reactions. An intramolecular Zn-Barbier reaction with *Rieke zinc* has been performed [114] with the rather complicated molecule as given below:

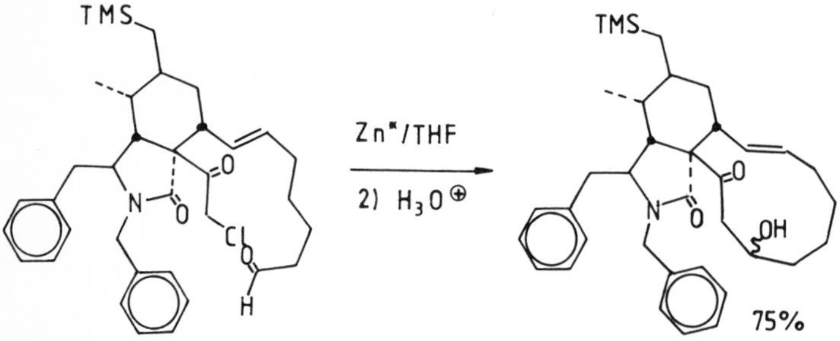

Zn^{*}/THF

2) $H_3O^{\oplus}$

75%

Treatment of the same aldehyde with SmI_2/THF results in cyclization to a single major isomer (46% was isolated).

Zinc/Graphite. Transition metal lamellar compounds of graphite can be prepared by utilizing the reducing properties of potassium–graphite (C_8K) towards transition metal salts [115] (see also [116]):

$$n\,C_8K + MX_n \xrightarrow{THF} C_{8n}M + n\,KX$$

MX_n is $Ti(i\text{-}PrO)_4$, $ZnCl_2$ and other salts

Zinc–graphite Zn/Gr prepared in this way, was used in several Reformatsky reactions as well as in Zn-Barbier reactions [116] in which allyl and crotyl bromides were involved. Products were obtained in excellent yields:

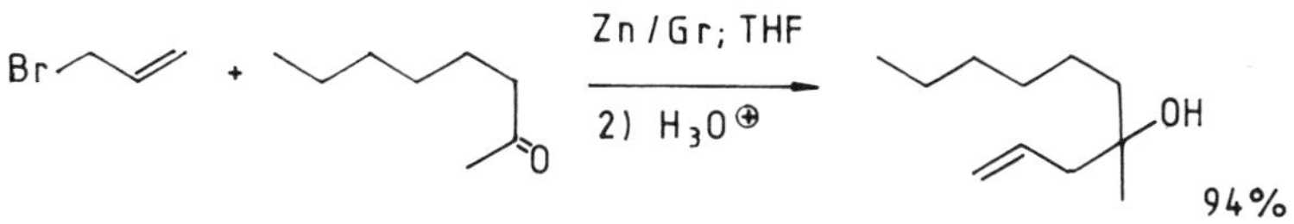

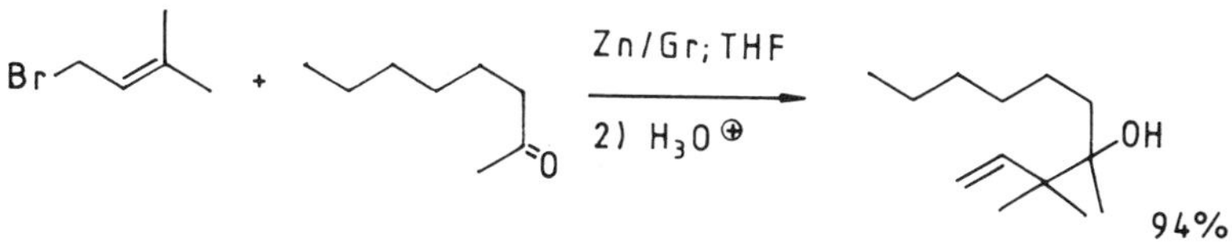

Zn/Gr; THF
2) $H_3O^{\oplus}$
94%

Zn/Gr; THF
2) $H_3O^{\oplus}$
94%

Zinc–Silver–Graphite reacts with ethyl dihaloacetates at low temperatures in a very efficient Reformatsky-type reaction to form α-halo-β-hydroxy alkanoates [117]:

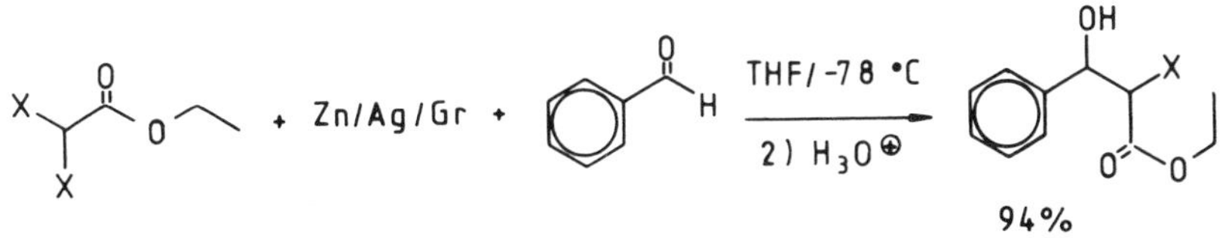

+ Zn/Ag/Gr +
THF/ −78 °C
2) $H_3O^{\oplus}$
94%

A recent publication on the Zn-Barbier reaction of allylic halides and nitriles with Zn/Gr showed that often the best yields were obtained with a 3:1 ratio of the allylic halide to the nitrile [118]:

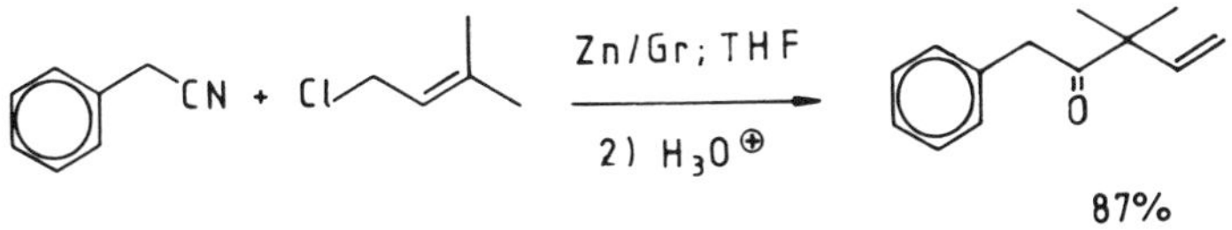

Zn/Gr; THF
2) $H_3O^{\oplus}$
87%

Some results, presented in Table 3.19, deserve special attention:

Table 3.21. Yields of unsaturated ketones through Barbier reactions of allylic halides and nitriles with Zn/graphite in THF [118]

Allylic halide	Nitrile	Product	Yield(%)
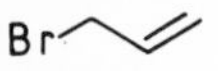 Br	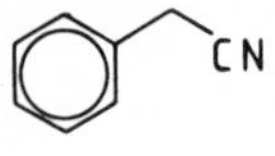 CN	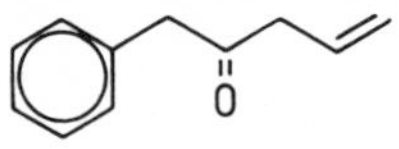 O	0
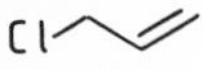 Cl	″	″	0
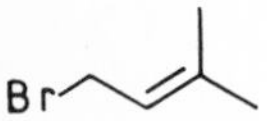 Br	″	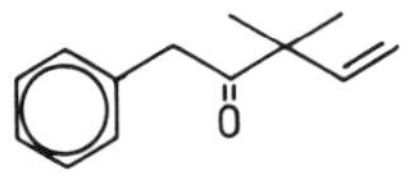 O	83
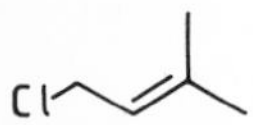 Cl	″	″	87
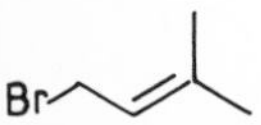 Br	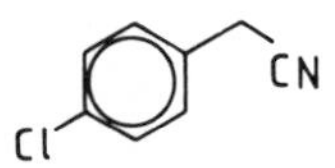 Cl CN	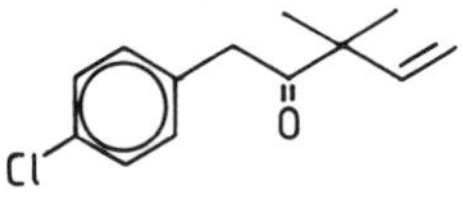 Cl O	80
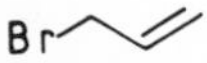 Br	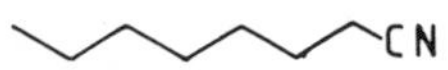 CN	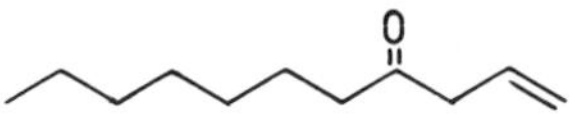 O	64
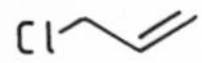 Cl	″	″	61

Allyl bromide or chloride did not yield any of the expected product in reactions with benzonitrile. However, with cyanoheptane the yields with these two halides were satisfactory.

The reactivities of other bromo and chloro derivatives differed insignificantly for most nitriles studied. This behaviour seems to indicate that the organozinc derivative is activated by the graphite support and suggests that the reaction with the nitrile occurs at the surface of the carbonaceous matrix rather than in the solution.

3.4.5 Conclusion

It is the intention in Sect. 3.4 to demonstrate the explosive development of Zn-Barbier reactions in the last fifteen to twenty years.

References quoted are mainly from the recent literature and there is little doubt that in the time between the finishing of this manuscript and its publishing a number of important new contributions will have been made.

Although in the main, the Saytzeff reactions of allylic halides have been investigated it is to be expected that new developments will be aimed at the use of other, less reactive halides.

The inclusion in this group of 1-bromoadamantane (p. 117, ref. [100b]), with which successful Barbier-type reactions have been reported mainly with lithium in Sect. 3.3.6, indicates that other less reactive halides will soon follow.

The reported use of water and alcohols as solvents for Zn-Barbier reactions is an astounding development and will stimulate more research in the use of water as solvent in organic synthesis, also in view of increasingly rigid environmental laws.

Further researches with these solvents will also lead to a better understanding of the mechanism of the (Zn-)Barbier reaction.

3.5 Other Metals in Barbier-Type Reactions

After having started this monograph with one-step carbon–carbon bond formation reactions in which magnesium was the metallic reagent (chap. 2) and then having progressed on to the use of lithium and zinc, some attention will now be focussed on the application of numerous other metals in Barbier-type reactions.

Among these metals tin and aluminium are predominant although their use seems restricted to allylic halides.

The first publication of what was named *the Grignard type reaction just using metallic tin* reported the following one-step procedure [119]:

$$X\text{-}CH_2CH{=}CH\text{-}R + Sn + R'C(O)R'' \xrightarrow[2)\ H_3O^{\oplus}]{THF} CH_2{=}CH\text{-}CH(R)\text{-}C(OH)R'R''$$

X = Br or I

It was found that

a. a wide variety of carbonyl compounds reacted smoothly,
b. the reaction was carried out simply by stirring the mixture in THF under essentially neutral conditions,
c. the Wurtz-coupling product was not formed, and
d. both allyl iodide and bromide could be used.

Since preformed diallyltin dihalide [88, 120] reacts smoothly with the carbonyl compounds investigated it was assumed that this organometallic reagent is initially formed in this one-step procedure.

A few years later it was found [89] that the presence of water, together with catalytic amounts of hydrobromic acid, in the reaction mixture also gave *satisfactory* results.

In a typical procedure hexanal, allyl bromide and commercially available tin powder were vigourously stirred in a 1:1 mixture of diethyl ether, water and 5% hydrobromic acid at room temperature. The expected product was obtained in 70% yield:

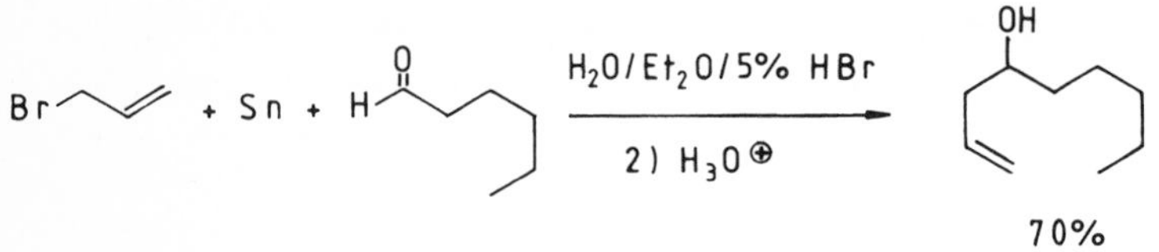

Br + Sn + H(C=O) → $H_2O/Et_2O/5\%$ HBr; 2) $H_3O^{\oplus}$ → OH 70%

When studying the same reaction with a variety of allylic bromides it was found that the addition of metallic aluminium *dramatically improved the yield* (aluminium chloride, aluminium oxide and metallic zinc were ineffective).

The following example demonstrates this new reaction:

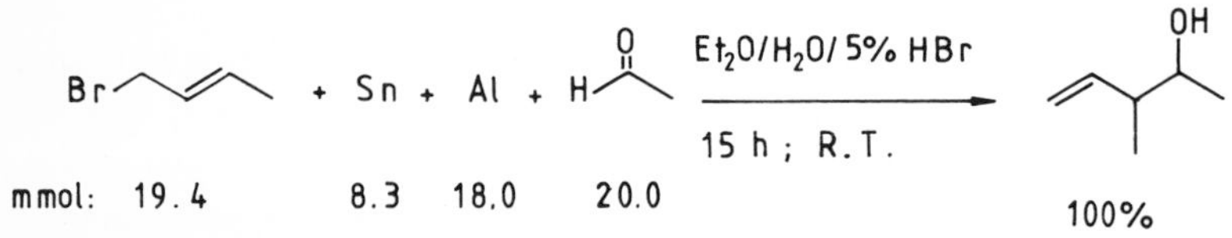

Br + Sn + Al + H(C=O) → $Et_2O/H_2O/5\%$ HBr; 15 h ; R.T. → OH 100%

mmol: 19.4 8.3 18.0 20.0

Later, the same group announced [121] *a facile one-pot synthesis of bromo homoallyl alcohols and of* 1,3-*keto acetates via allyltin intermediates.* A variety of carbonyl compounds were applied in Sn-Barbier reactions with 2,3-dibromoprop-1-ene as well as with 2-acetoxy-3-bromoprop-1-ene.

The following examples demonstrate the versatility of the reaction:

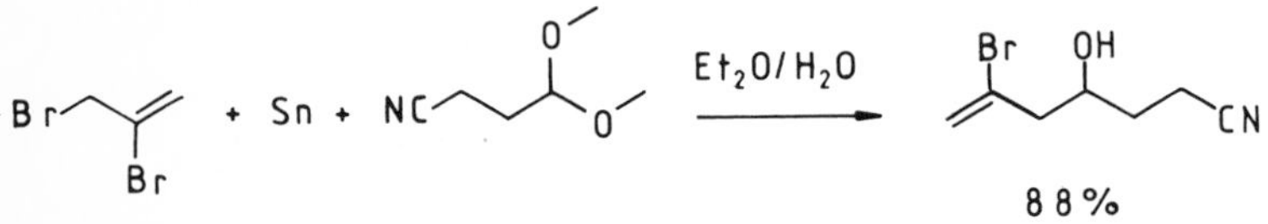

Br(Br) + Sn + NC(O)(O) → Et_2O/H_2O → Br OH CN 88%

and

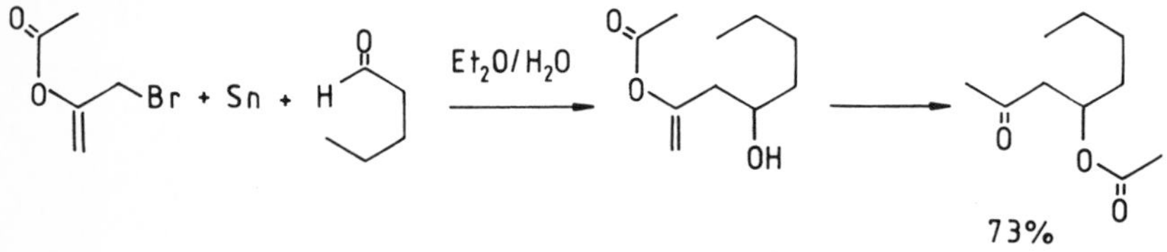

O O Br + Sn + H(O) → Et_2O/H_2O → O O OH → O O O 73%

In a newly developed *electrochemical Grignard-type allylation of carbonyl compounds* allyltin reagents are said to be *recycled* [122].

In a typical example, using a platinum foil cathode, benzaldehyde, allyl bromide, tin powder and cyclohexene, dissolved in a 4:1 mixture of methanol and methanoic acid, were electrolyzed at 50–55 °C, affording the expected product in 91% yield:

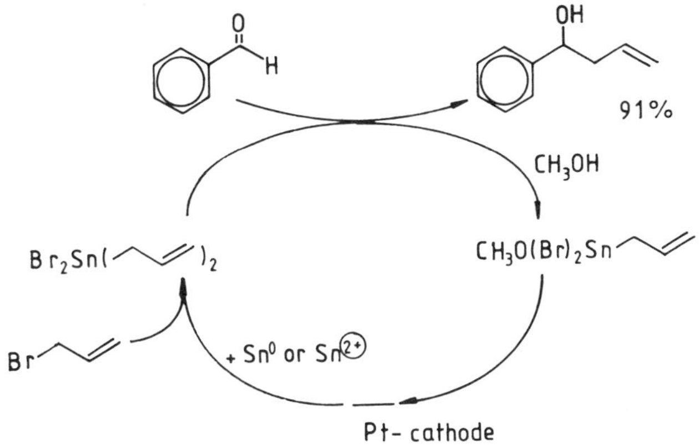

91%
CH_3OH
$Br_2Sn(CH_2CH{=}CH_2)_2$
$CH_3O(Br)_2SnCH_2CH{=}CH_2$
Br
+ Sn^0 or Sn^{2+}
Pt- cathode

In general, yields with various carbonyl compounds were very satisfactory; worth particular mention is that hydroxy-substituted substrates also gave excellent results:

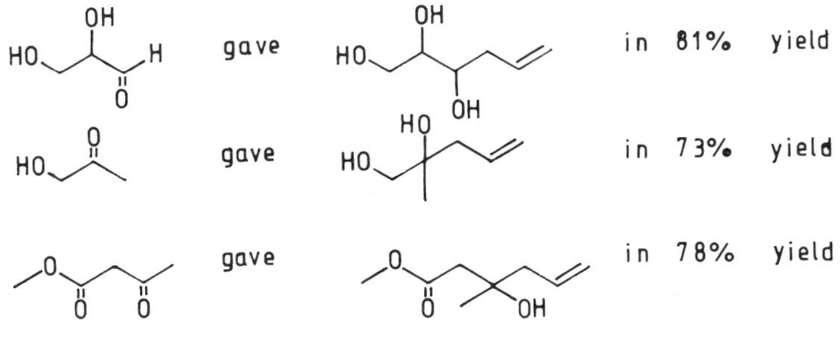

gave in 81% yield

gave in 73% yield

gave in 78% yield

It was later found [123] that 'recycling' the allyltin reagents could also be effected with the aid of metallic aluminium. This metal is less electronegative than tin and so in principle reduces di- and tetravalent-tin to metallic tin.

Furthermore, it was found that allyl chloride could be used in these reactions instead of the more expensive bromide.

To generate and recycle an active zero-valent tin, a combination of 2 equivalents of aluminium powder and 0.1 equivalent of tin (II) was used in various solvents and solvent mixtures such as DME-methanol (1:1), 2-methoxy-1-ethanol. However water and methanoic acid had always to be added.

The best yields were obtained when the aluminium/aldehyde ratio was 1:1. The results suggest that zero-valent tin alone is not sufficient to give a satisfactory reaction and that zero-valent tin on a metallic aluminium surface is responsible for the efficient allylation with allyl chloride.

The following scheme represents the recycling procedure:

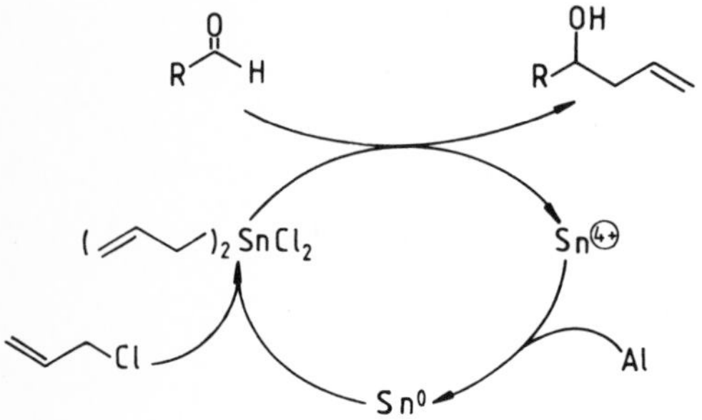

As was the case in the previously discussed electrochemical process, hydroxy-substituted aldehydes gave excellent yields of the expected products; glyceraldehyde e.g. gave 82% yield of 1, 2, 3-trihydroxy-5-hexene:

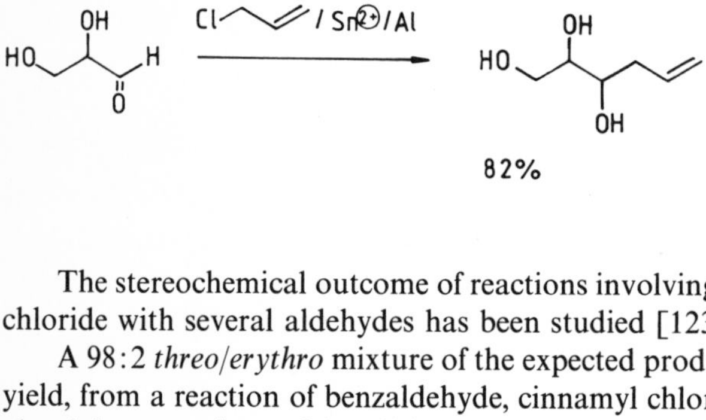

The stereochemical outcome of reactions involving Sn[II]/Al and cinnamyl chloride with several aldehydes has been studied [123, 125].

A 98:2 *threo/erythro* mixture of the expected product was obtained, in 82% yield, from a reaction of benzaldehyde, cinnamyl chloride and a 2:1 mixture of aluminium powder and Sn_2Cl in THF/water:

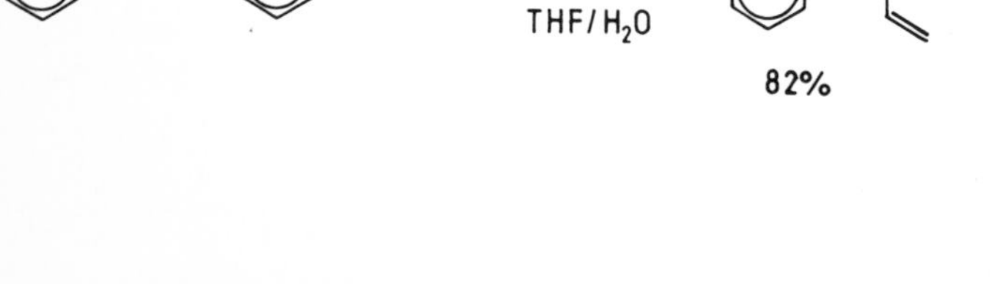

Sonication has also been applied to Barbier-type reactions with zinc and tin.

Allyl bromide reacted with tin in the presence of a mixture of benzaldehyde and acetophenone in a 5:1 H_2O/THF solvent mixture to give 90% yield of the aldehyde addition product whereas the ketone did not react at all under these competitive conditions [92]:

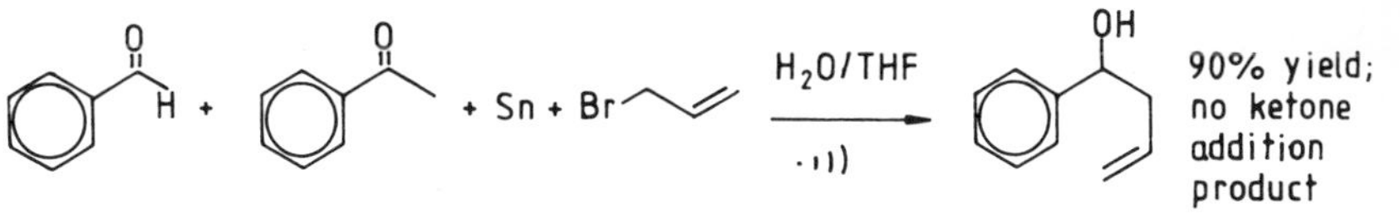

H + + Sn + Br

H_2O/THF

·))

OH

90% yield; no ketone addition product

As previously observed, the presence of hydroxy-substituents in the substrates did not influence the efficiency of the reaction [93].

The electroreductive Barbier-type allylation was applied recently to imines [126] in the following overall reaction:

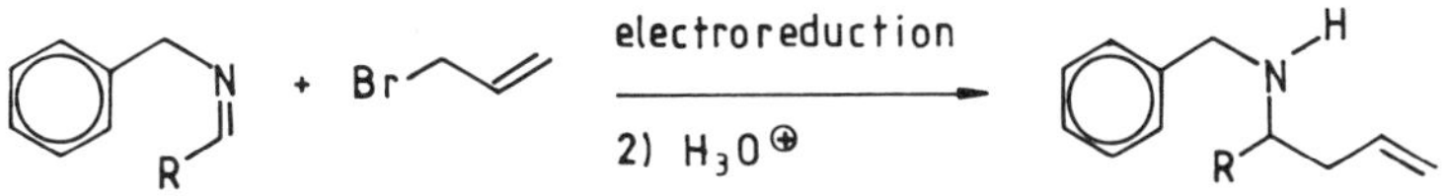

N R + Br

electroreduction

2) $H_3O^{\oplus}$

N H R

With platinum as the cathode the most efficient anode-metal was zinc or aluminium. For the second metal redox couple (see Scheme below) various metals were used, Pb[II] bromide and Bi[III] chloride giving the best yields. Tin and zinc salts were less efficient:

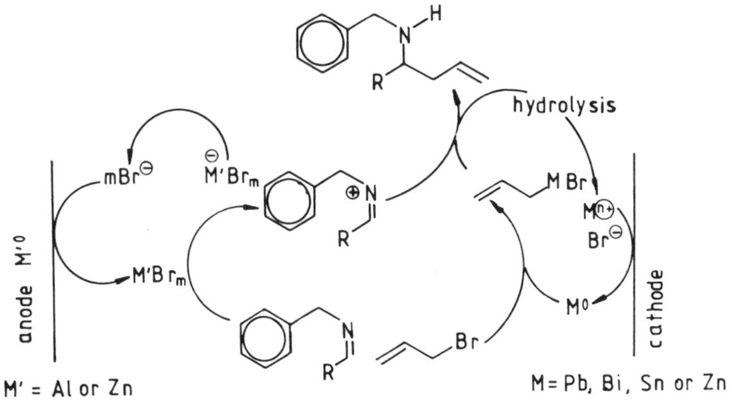

N H R

hydrolysis

mBr⊖

⊖ M'Br$_m$

⊕N R

M Br

M(n+)

Br⊖

anode M'0

M'Br$_m$

cathode

M0

N R

Br

M' = Al or Zn

M = Pb, Bi, Sn or Zn

With R being various aromatic groups or cyclohexyl, yields were as high as 82%–98%.

Shortly before the discovery of the efficiency of Pb[II] in the electroreduction the first *lead-promoted Barbier-type reaction* had been published [127].

The procedure was as follows: propargyl bromide was mixed with benzaldehyde, tetra-*n*-butylammonium bromide and trimethylchlorosilane (TMSCl) in DMF. Whilst stirring, a lead plate was immersed in the reaction mixture until most of the aldehyde was consumed. After acidic hydrolysis a mixture of two products was obtained in 95% yield:

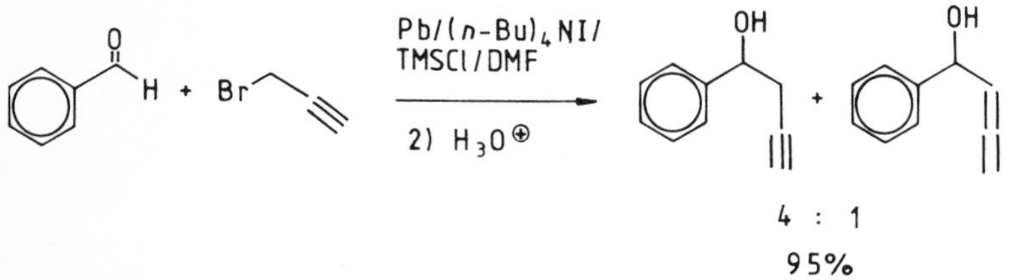

Pb/(n-Bu)$_4$NI/ TMSCl/DMF; 2) $H_3O^{\oplus}$ — 4 : 1, 95%

Various aldehydes were studied among which were furfuraldehyde (83% yield) and heptanal (90% yield).

It was suggested that the lead-promoted reaction involves the formation of an active divalent organolead reagent with the aid of the ammonium bromide, present in solution. A possible structure for the intermediate could be the following:

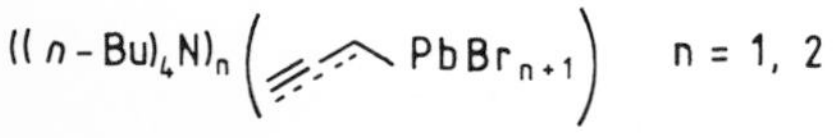

$((n\text{-Bu})_4N)_n\left(\ldots PbBr_{n+1}\right)$ $n = 1, 2$

This salt would react with the aldehyde to give the expected products with liberation of Pb[II] bromide.

Also Pb-Barbier reactions with allylic bromides were soon discovered [128].

Allylation of acetals was made possible [129] by the use of Pb[II]/Al in the in situ reaction with allyl bromide and acetals. Cationic intermediates are supposed to react with allyllead reagents formed from zero-valent lead as represented in the following scheme:

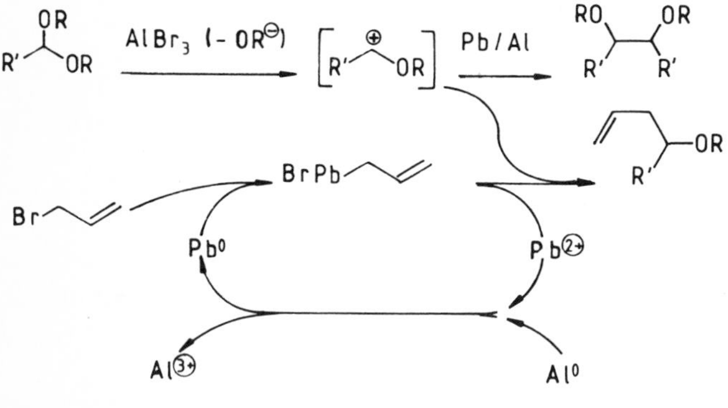

AlBr$_3$ (- OR$^{\ominus}$); Pb/Al; BrPb; Br; Pb^0; Pb^{2+}; Al^{3+}; Al^0

The *first example of the Grignard-type allylation of aldehydes just using metallic bismuth and allyl halides* was reported only a few years ago [130].

Among group VB elements bismuth metal is cheaper and less toxic than arsenic or antimony, and, in the opinion of the authors

can be expected to play some role in organic synthesis according to its enhanced metallic character.

Stirring metallic bismuth with a mixture of an allyl halide and an aldehyde in DMF at room temperature did indeed give high yields of the expected homoallylic alcohols:

Br + H + Bi — DMF/R.T./12h, 2) $H_3O^{\oplus}$ → OH 80%

With benzaldehyde yields were almost quantitative.

A follow-up to this work led to the discovery that Bi[III], combined with metals such as zinc, iron and aluminium also gave excellent yields of the expected products [131–133].

Very recently the combination Bi[III]/Al was applied in a THF/water mixture for a one-step carbon-carbon bond formation reaction in the synthesis of amines [134]. Benzyl bromide as well as iodomethane were used as the halides:

N N N N + Br — Bi^{3+}/Al, THF/H_2O → N 79%

With iodomethane this reaction gave 75% yield of the same type of product (*N*, *N*-diphenylaminoethane).

N

Only very recently another *first example* of a Barbier-type reaction with a 'new' metal was announced [135].

This time it was a *first example of Barbier-Grignard-type allylation of aldehydes using metallic antimony.*

In several reactions the yields were excellent.

Allylation with metallic manganese (with particle size of 10 μ) [136] in THF also gave fair yields with various carbonyl compounds (benzaldehyde, 82%; acetophenone, 76%; 4-*tert*-butyl-cyclohexanone, 89%; cinnamaldehyde, 78% yield). For these reactions 10 mol% iodine had to be added to the metal.

Several years later it was stated [137] by a different group of researchers that *commercial micronized manganese powders must be avoided, since they contain large amounts of oxides.... moreover the manganese had to be activated by addition of one equivalent of iodine.*

These authors used a *commercial coarse-ground manganese, which is easily available and cheap.* The solvent of choice for Barbier-type reactions with this metal was ethyl ethanoate. Yields of reactions with aliphatic ketones were excellent. In order to get such good yields with aliphatic aldehydes 10% $ZnCl_2$, $CdCl_2$, or $HgCl_2$ had to be added.

Another important result of this work was that addition of $ZnCl_2$ to the reaction mixture allowed the reaction to be performed in THF which gave excellent results with both aldehydes and ketones.

Cerium has also been applied in a one-step synthetic procedure [138]; as was calcium which was used with iodoperfluoroalkanes [139].

Copper was used in carbon-carbon bond formation reactions for the first time in 1904 [140] in the well-known Ullmann reaction which involved aromatic halides. Very recent work has reported the use of 'activated' copper (Cu*) in homo-coupling and cyclization reactions of α, ω-dihaloalkanes [141].

Iron was among the metals, used in the electrosynthesis of a wide range of alcohols from organic halides and ketones or aldehydes (see Sect. 2.4 and 3.4 p. 121) [112].

Electrolysis of tetrachloromethane and benzaldehyde in DMF (and a small amount of supporting electrolyte) with a sacrificial iron electrode yielded 60% of 1,1,1-trichloro-2-phenyl-2-ethanol:

CCl_4 + Fe (electrode) + PhCHO $\xrightarrow[\text{2) } H_3O^{\oplus}]{\text{DMF/electrolyte}}$ PhCH(OH)CCl_3 60%

3.6 Summary

This chapter has demonstrated that magnesium has met with 'great competition' as the metal of choice in one-step, Barbier-type reactions.

As noted in Chapter Two it was not until 1970 that reports of such reactions with metals, other than magnesium, could be found in the chemical literature.

Since then, metals such as lithium (Sect. 3.3) and zinc (Sect. 3.4) have started to draw the attention of synthetic organic chemists and the successes obtained with them have inspired researchers to extend this type of reaction even further.

The results of this research were dealt with in Sect. 3.5.

It remains to be seen to what extent these new developments will be applied in everyday laboratory practice.

There is little doubt, however, that applications on an industrial scale will make such reactions more and more attractive. The first patent in this field was issued in 1930 (see chap. 2; [42]) and many have followed.

As mentioned several times in the present chapter, metals such as zinc (Sect. 3.4) and tin (Sect. 3.5) allow aqueous solvents to be used and this may prove to be of crucial importance for industrial application; particularly in light of increasing environmental requirements by local, national and/or international authorities.

It has already been stated that the discovery of new reactions and of new reaction conditions is presently so rapid that they can hardly be dealt with in one single monograph.

It is therefore to be hoped that additional specialized publications will continue to appear to fill the gap that separates the laboratory researcher from the flow of information that is presenting itself daily.

3.7 References

1. Wurtz A (1854) Ann Chim et Phys [3], 44: 129
2. Wurtz A (1855) Ann Chim et Phys [3], 44: 275
3. Tollens B and Fittig R (1864) Ann 131: 303
4. Brigel G (1865) Ann 135: 171
5. Klinger H (1883) Ber 16: 1994
6. Barbier Ph (1899) Compt Rend 128: 110
7. Frey H (1895) Ber 28: 2514
8. Schorigin P (1907) Ber 40: 3111
9. Freund A (1861) Ann 118: 1
10. Grignard V (1900) Compt Rend 130: 1322
11. Grignard V (1901) Ann Chim et Phys 24: 433
12. Danishefski S and Dumas DJ, J Chem Soc Chem Chem Comm 1968, 1287
13. Mirrington RN and Schmalzl KJ (1972) J Org Chem 37: 2871
14. Pearce PJ, Richards DH and Scilly NF, J Chem Soc Chem Comm 1970: 1160
15. Pearce PJ, Richards DH and Scilly NF, J Chem Soc Perkin Trans I 1972: 1655
16. Einhorn J and Luche J-L (1986) Tetrahedron Lett 27: 501
17. Leroux Y, Bull Soc Chim France 1968: 359
18. Hamon DPG and Sinclair RW, J Chem Soc Chem Comm 1968: 890
19. Scilly NF, J Chem Soc Chem Comm 1973: 160
20. Katzenellenbogen JA and Lennox RS, Tetrahedron Lett 1972: 1471
21. Katzenellenbogen JA and Lennox RS (1973) J Org Chem 38: 326
22. Larcheveque M, Debal A and Cuvigny Th (1975) J Organometal Chem 87: 25
23. Cameron GG and Milton AJS, J Chem Soc Perkin Trans II 1976: 378
24. Bauer P and Molle G (1978) Tetrahedron Lett 48: 4853
25. Molle G and Bauer P (1982) J Amer Chem Soc 104: 3481
26. Lomas JS (1984) Nouv J Chimie 8: 365
27. Mason TJ, Lorrimer JP (1988) Sonochemistry: Theory, Applications and Uses of Ultrasound in Chemistry Ellis Horwood Chichester p 75

28. Luche J-L (1990) In: Mason TJ (Ed) Advances in Sonochemistry Vol 1 JAI Press Greenwich, p 119
29. Luche J-L and Damiano J-C (1980) J Amer Chem Soc 102: 7927
30. Trost BM and Coppola BP (1982) J Amer Chem Soc 104: 6879
31. Snowden RG and Sonay Ph (1984) J Org Chem 49: 1464
32. Uyehara T, Yamada J, Ogata K and Kato T (1985) Bull Chem Soc Japan 58: 211
33. Ihara M, Katogi M, Fukumoto K and Kametani T, J Chem Soc Chem Comm 1987: 721
34. de Souza-Barboza JC, Luche J-L and Pétrier C (1987) Tetrahedron Lett 28: 2013
35. de Souza-Barboza JC, Pétrier C and Luche J-L (1988) J Org Chem 53: 1212
36. Choi H, Pinkerton AA and Fry JL, J Chem Soc Chem Comm 1987: 225
37. Shih N-Y, Mangiaracina P, Green MJ and Ganguly AK (1989) Tetrahedron Lett 30: 5563
38. Kihara M, Kashimoto M, Kobayashi Y and Kobayashi S (1990) Tetrahedron Lett 37: 5347
39. Frankland E (1849) Ann 71: 171
40. Frankland E (1849) Ann 71: 213
41. Saytzeff A (1870) Z f Chemie 13: Neue Folge 6: 104
42. Wagner G and Saytzeff A (1875) Ann 175: 351
43. Reformatsky A (1887) Ber 20: 1210
44. Simmons HE and Smith RD (1958) J Amer Chem Soc 80: 5323
45. Simmons HE and Smith RD (1959) J Amer Chem Soc 81: 4256
46. Furukawa J and Kawabata N (1974) Adv Organometal Chem 12: 83
47. Nützel K (1973) In Methoden der Organischen Chemie Houben-Weyl Vol 13/2a Metallorganische Verbindungen Georg Thieme, Stuttgart
48. Boersma J (1982) In Comprehensive Organometallic Chemistry (Eds Wilkinson G, Stone FGA and Abel EW) Pergamon Oxford
49. Pébal L (1861) Ann 118: 22
50. Rieth R and Beilstein F (1863) Ann 126: 241
51. Gladstone JH and Treibe A (1876) J Chem Soc 26: 445
52. See [47] p. 570
53. Rieke RD, Uhm SJ and Hudnall PM, J Chem Soc Chem Comm 1973: 269
54. Rieke RD and Uhm SJ, Synthesis 1975: 452
55. Rieke RD (1977) Acc Chem Res 10: 301
56. Rieke RD (1989) Science 246: 1260
57. Csuk R, Glänzer BI and Fürstner A (1988) Adv Organometall Chem 28: 85
58a. Han BH and Boudjouk P (1982) J Org Chem 47: 5030
58b. Kitazume T and Ishikawa N, Chem Lett 1981: 1679
59. Zeile K and Meyer H (1942) Ber 75: 356
60. Henbest HB, Jones ERH and Walls IMS, J Chem Soc 1949: 2696
61. Henbest HB, Jones ERH and Walls IMS, J Chem Soc 1950: 3646
62. Golse R and Le Van Thoi (1950) Compt Rend 230: 210
63. Keskin H, Miller RE and Nord FF (1951) J Org Chem 16: 199
64. Wotiz JH (1951) J Amer Chem Soc 73: 693
65. Entschel R, Eugster CH and Karrer P (1956) Helv Chim Acta 39: 686
66. Eugster CH, Linner E, Trivedi AH and Karrer P (1956) Helv Chim Acta 39: 690
67. Wotiz JH and Mancuso DE (1957) J Org Chem 22: 207
68. Bailey AS, Kendall VG, Lumb PB, Smith JC and Walker CH, J Chem Soc 1957: 3027
69. Bohlmann F, Herbst P and Gleinig H (1961) Ber 94: 948
70. Hanzawa Y, Inazawa K, Kon A, Aoki H and Kobayashi Y (1987) Tetrahedron Lett 28: 659
71. Blancou H, Morreau P and Commeyras A, J Chem Soc Chem Comm 1976: 885
72. Miller Jr WT, Bergmann E and Fainberg AH (1957) J Amer Chem Soc 79: 4159
72a. Wakselman C and Tordeux M, Bull Soc Chim France 1986: 868
73. Kitazume T and Ishikawa N (1985) J Amer Chem Soc 107: 5186
74. Fujita M, Morita T and Hiyama T (1986) Tetrahedron Lett 27: 2135
75. Fujita M, Hiyama T and Kondo K (1986) Tetrahedron Lett 27: 1239
76. Fujita M and Hiyama T (1986) Tetrahedron Lett 27: 3655
77. O'Reilly NJ, Maruta M and Ishikawa N, Chem Lett 1984: 517
78. Francèse C, Tordeux M and Wakselman C, J Chem Soc Chem Comm (1987): 642
79. Francèse C, Tordeux M and Wakselman C (1988) Tetrahedron Lett 29: 1029
80. Motherwell WB, J Chem Soc Chem Comm 1973: 935
80a. Hodge P and Khan MN, J Chem Soc Chem Comm 1975: 809
81. Corey EJ and Pyne SG (1983) Tetrahedron Lett 24: 2821

82. Gawronski JK (1984) Tetrahedron Lett 25: 2605
83. Palomo C, Aizpurua JM, López C and Aurrekoetxea N (1990) Tetrahedron Lett 31: 2205
84. Palomo C, Aizpurua JM and Aurrekoetxea N (1990) Tetrahedron Lett 31: 2209
85. Palomo C, Aizpurua JM, López MC, Aurrekoetxea N and Oiarbide M (1990) Tetrahedron Lett 31: 6425
86. Picotin G and Miginiac Ph (1987) Tetrahedron Lett 28: 4551
86a. Smith CL, Arnett J and Ezike J, J Chem Soc Chem Comm 1980: 653
87. Killinger TA, Boughton NA, Runge TA and Wolinsky J (1977) J Organometal Chem 124: 131
88. Sisido K, Takeda Y and Kinugawa Z (1961) J Amer Chem Soc 83: 538
89. Nokami J, Otera J, Sudo T and Okawara R (1983) Organometall 2: 191
90. Pétrier C and Luche J-L (1985) J Org Chem 50: 910
91. Kerdesky FAJ, Ardecky RJ, Lakshminkantham MV and Cava MP (1981) J Amer Chem Soc 103: 1992
92. Pétrier C, Einhorn J and Luche J-L (1985) Tetrahedron Lett 26: 1449
93. Einhorn C and Luche J-L (1987) J Organometal Chem 322: 177
94. Wilson SR and Guazzaroni ME (1989) J Org Chem 54: 3087
95. Shono T, Nishiguchi I and Sasaki M (1978) J Amer Chem Soc 100: 4314
96. Shono T, Hamaguchi H, Sasaki M, Fujita S and Nagami K (1983) J Org Chem 48: 1621
97. Knochel P and Normant J-F (1984) Tetrahedron Lett 25: 1475
98. Knochel P and Normant J-F (1984) Tetrahedron Lett 25: 4383
99. Knochel P and Normant J-F (1985) Tetrahedron Lett 26: 425
100. Pietrusiewicz KM and Zoblocka M (1988) Tetrahedron Lett 29: 937
100a. Castedo L, Mascareñas JL, Mouriño A and Sarandeses LA (1988) Tetrahedron Lett 29: 1203
100b. Ohno M, Ishikawa K and Eguchi S (1988) J Org Chem 53: 1285
101. Li C-S, Cheng C-H, Cheng S-S and Shaw J-S, J Chem Soc Chem Comm 1990: 1774
102. Blaise EE (1904) Compt Rend 138: 284
103. Denis JM, Girard C and Conia JM, Synthesis 1972: 549
104. Rousseau G and Conia JM (1981) Tetrahedron Lett 22: 649
105. Nazarov IN and Zaretskaya II (1957) Zhur Obshchei Khim 27: 624 (1957) Chem Abstr 51: 16316f (1957)
106. Rousseau G and Drouin J (1983) Tetrahedron 39: 2307
107. Han BH and Boudjouk P (1982) J Org Chem 47: 752
108. Joshi NN and Hoffmann HMR (1986) Tetrahedron Lett 26: 687
109. Mortimore M and Kocienski P (1988) Tetrahedron Lett 29: 3357
110. Lenoir D, Synthesis 1989: 883
111a Maruoka K, Hashimoto S, Kitagawa Y, Yamamoto H and Nozaki H (1977) J Amer Chem Soc 99: 7705
111b. Maruoka K, Hashimoto S, Kitagawa Y, Yamamoto H and Nozaki K (1980) Bull Soc Chem Japan 53: 3301
112. Sibille S, d'Incan E, Leport L and Périchon J (1986) Tetrahedron Lett 27: 3129
113. Tanaka H, Nakahara T, Dhimane H and Torii S (1989) Tetrahedron Lett 30: 4161
114. Vedejs E and Ahmad S (1988) Tetrahedron Lett 29: 2291
115. Braga D, Ripamonti A, Savoia D, Trombini C and Umani-ronchi A, J Chem Soc Chem Comm 1978: 927
116. Boldrini GP, Savoia D, Tagliavini A, Trombini C and Umani-Ronchi A (1983) J Org Chem 48: 4108
117. Fürstner A (1987) J Organometal Chem 336: C33
118. Marceau P, Gautreau L and Béguin F (1991) J Organometal Chem 403: 21
119. Mukaiyama T and Harada T, Chem Lett 1981: 1527
120. Vijayaraghavan KVJ (1945) Indian Chem Soc 22 135; Chem Abstr 40: 2787 (1946)
121. Mandai T, Nokamı J, Yano T, Yoshinaga Y and Otera J (1984) J Org Chem 49: 172
122. Uneyama K, Matsuda H and Torii S (1984) Tetrahedron Lett 25: 6017
123. Uneyama K, Kamaki N, Moriya A and Torii S (1985) J Org Chem 50: 5396
124. Uneyama K, Nanbu H, and Torii S (1986) Tetrahedron Lett 27: 2395
125. Coxson JM, Van Eyk SJ and Steel PJ (1985) Tetrahedron Lett 26: 6121
126. Tanaka H, Nakahara T, Dhimane H and Torii S (1989) Tetrahedron Lett 30: 1461
127. Tanaka H, Hamatani T, Yamashita S and Torii S, Chem Lett 1986: 1461
128. Tanaka H, Hamatani T, Yamashita S, Ikamoto Y and Torii S, Chem Lett 1986: 1611
129. Tanaka H, Yamashita S, Ikemoto Y and Torii S (1988) Tetrahedron Lett 29: 1721
130. Wada M and Akiba K (1985) Tetrahedron Lett 26: 4211

131. Wada M, Ohki K and Akiba K (1986) Tetrahedron Lett 27: 4771
132. Wada M, Ohki H and Akiba K, J Chem Soc Chem Comm 1987: 708
133. Wada M, Ohki H and Akiba K (1990) Bull Chem Soc Japan 63: 1738
134a. Katritzky AR, Shobana N and Harris PA (1991) Tetrahedron Lett 32: 4247
134b. Katritzky AR, Shobana N and Harris PA (1992) Organometallics 11: 1381
135. Butsugan Y, Ito H and Araki S (1987) Tetrahedron Lett 28 3707
136. Hiyama T, Sawahata M and Obayashi M, Chem Lett 1983: 1237
137. Cahiez G and Chavant P-Y (1989) Tetrahedron Lett 30: 7373
138. Inamoto T, Kusumoto T, Tawarayama Y, Yasushi Y, Mita T, Hatanaka Y, and Yokoyama M (1984) J Org Chem 49: 3907
139. Santini G, Le Blanc M and Riess JG (1977) J Organometal Chem 140: 1
140. Ullmann F (1904) Ann 332: 38
141. Ginah FO, Donovan Jr TA, Suchan SD, Pfennig DR and Ebert GW (1990) J Org Chem 55: 584

4 The Mechanism of the Barbier Reaction

4.1 Introduction

One of the crucial factors in the Barbier reaction, just as in the reaction to form the Grignard reagent, is the interaction between the organo halide and the metal.

With the simultaneous presence of three reactants, i.e. the organo halide, substrate and metal, the mechanism of the Barbier reaction could be said to be even more difficult to elucidate, since interaction with the metal could take place consecutively with each reagent or with both at the same time.

To make matters worse, a fourth organic 'reagent' present in the reaction mixture must be considered: the solvent or solvent-mixture. How does this interfere with the reagent processes and interactions?

However, only a relatively small number of publications have appeared on the mechanistic aspects of these reactions. This is in sharp contrast to other catalytic processes on metal surfaces which have drawn tremendous attention in industrial research.

Fortunately, thanks to the development of more sophisticated techniques in the last two or three decades, as well as to a growing interest in this field of organometallic chemistry, several promising, systematic studies have appeared on the formation mechanism of the Grignard reagent.

And very recently, systematic research on the mechanism of the Barbier reaction has also been published.

With an increasing understanding of this heterogeneous reaction these investigations have also been extended to the formation reaction of organo-lithium, -zinc and -aluminium reagents.

This chapter on the mechanism of the Barbier reaction discusses the following:

1. the interaction of metals and organic halides (Sect. 4.2);
2. the Grignard-type reaction (Sect. 4.3);
3. the Barbier-type reaction (Sect. 4.4);

4.2 Mechanism of the Interaction of Metals with Organo Halides

4.2.1 Early Work by Polanyi et al.; 1934

In the early 1930s 'sodium-flame' reactions were studied [1] in which sodium vapour was brought into contact with organo halides in a carrier-gas leading to the formation of radicals:

$$Na + R\text{-}Hal = Na\text{-}Hal + R\cdot$$

4.2.2 'High Temperature Species'; 1967

Almost forty years later use was made of new techniques [2] to work with 'high-temperature' species in which carbon arc vapours, with C_3 as the major reactant, were condensed in a hydrogen matrix at low temperatures to react with unsaturated hydrocarbons such as *cis*- and *trans*-2-butene.

Then methylenecyclopropane was synthesized by reaction of potassium vapour with 1-iodo-2-iodomethyl-2-propene in a form of intramolecular Wurtz reaction [3]:

I I K-vapour

4.2.3 Vapourization of Metals and Codeposition with Reactants

Further fields of study were opened up when techniques were developed for the deposition of vapours of less volatile metals on cold surfaces, allowing the formation of pure metal-layers of varying thickness [4].

Ground-State Magnesium. This led to investigations [5] of chemical reactions with ground-state magnesium, ^1S Mg in 1972.

Codeposition of this metal with a large excess of water (> 100:1) gave hydrogen in more than 90% yield after warming up:

$$Mg + H_2O \longrightarrow H_2$$

With ammonia the yield of hydrogen was less than 3%:

$$Mg + NH_3 \longrightarrow H_2$$

When magnesium was codeposited with organo halides, solvent-free Grignard reagents were obtained. The yields (determined by hydrolysis) for each of the following halides were:

1-iodopropane	70%;	2-iodopropane	55%;
1,1-dimethyl-1-bromoethane	5%;	chlorobenzene	58%;
bromoethene	78%.		

When first 1-bromopropane was codeposited with magnesium at low temperature, followed by bromomethane at the same low temperature, hydrolysis yielded methane exclusively:

$$C_3H_7Br + Mg \xrightarrow{-110\ °C} + CH_3Br \xrightarrow[2)\ H_3O^{\oplus}]{-110\ °C \rightarrow R.T.} CH_4$$

When bromomethane was added after the mixture of magnesium and 1-bromopropane had been allowed to warm up propane was the only product on hydrolysis:

$$C_3H_7Br + Mg \xrightarrow{-110\ °C \rightarrow R.T.} + CH_3Br \xrightarrow[2)\ H_3O^{\oplus}]{-110\ °C \rightarrow R.T.} C_3H_8$$

The following interpretation of these observations was given:

Consequently it is proposed that in the black matrix the magnesium atoms are weakly bound to the halogen atom of the alkyl halide thus limiting its movement and selfcondensation to make a film; magnesium atoms bound in this manner are readily transferred to other alkyl halides and since methyl halides are more reactive than propyl, the former reacts preferentially.

With certain reactants the solvent-free Grignard reagents, formed in these experiments, react differently from the Grignard compounds formed under more common conditions.

Zinc Vapour. In that same period others [6] investigated reactions with zinc atoms. Non-solvated fluoroorganic zinc compounds were formed. The authors claim to

have found evidence that zinc atoms .... insert into C-I bonds and that the resultant organometallic compounds have vastly different properties than those generated by normal solution-phase techniques.

In an earlier paper [7] the same authors reported that calcium atoms reacted with perfluoro-2-butene at liquid nitrogen temperature to give perfluoro-2-butyne:

$$Ca + CF_3CF{=}CFCF_3 \xrightarrow{\text{liq. } N_2} CF_3C{\equiv}CCF_3 + CaF_2$$

The initial step in this reaction is the oxidative insertion of calcium to the carbon-fluorine bond (also named 'oxidative addition to metals'; see [8]), followed by rapid elimination of calcium fluoride.

Codeposition of Magnesium and a Ketone. A further interesting observation [9] by the same group was the interaction between metal atoms and a ketone, a reagent often present in a Barbier reaction mixture. Cocondensation of magnesium with cyclic ketones leads to both deoxygenation and dimerization:

Mg + O (cyclopentanone) —2) $H_3O^{\oplus}$→ cyclopentene + OH OH (1,1'-bicyclopentyl-1,1'-diol)

This is analogous to the standard procedure mentioned in many organic chemistry textbooks for this reaction which usually involves magnesium-amalgam [10].

Codeposition of Magnesium with HaloMethanes. In more recent work (see also [11] on products formed on codeposition of magnesium atoms and halomethanes in argon matrices) it was noted that rather high concentrations of magnesium were necessary for a successful experiment [12], suggesting that clusters of magnesium would be necessary for the oxidative addition reaction to occur.

Indeed two years later [13], UV-visible evidence was reported for a greater reactivity of Mg_2 and $(Mg)_x$ (with $x > 2$) than of Mg atoms in oxidative addition reactions with bromomethane.

The authors reasoned this to be due:

to more favorable reaction energetics as pointed by Dykstra [14]. .... *Clusters generally have a lower Ionization Potential than metal atoms, and our earlier work supported an electron transfer from Mg to CH_3Br to initiate the reaction. This primary step should be more favorable for a cluster, and in the process of ionization a considerably stronger Mg–Mg bond would form.*

Although cleavage of C–Br on Mg_2 would involve moving C and Br further apart in the final product, the initial bond-breaking act may be facilitated by a four-centered approach:

$(Mg\text{----}Mg)^{\oplus}$

$(C\text{----}Br)^{\ominus}$

The implications of this work with respect to elucidating the mechanism of Grignard reagent formation are that small clusters of magnesium may be removed from the surface and that cluster Grignard reagents $R(Mg)_xX$ may be stable species.

Further studies were undertaken regarding the reactivities of both calcium and magnesium with the four halomethanes [15]. Based on the bond-strength data for diatomic Mg–X vs C–X the CH_3X reactivity trend $CH_3I > CH_3F > CH_3Br > CH_3Cl$ is reasonable.

The authors were of the opinion that their results

.. reinforce our earlier proposal with theoretical support that Mg_2 can be more reactive than Mg atoms because a) Mg_2 can more easily transfer an electron to initiate the reaction process, and b) the logical product CH_3MgMgX should have a comparatively strong Mg–Mg bond.

The driving force for electron transfer may lie in the ion-pair stabilization that is gained in the solid matrix. Electron transfer strengthens M–M bonding, and a Coulombic ionic attraction between the product $M_x^+CH_3X^-$ species would be present.

This ionic stabilization could compensate for the mismatch in ionization potential of the metal species (about 6 eV) vs the electron affinity of CH_3X (about 1 eV).

Alternatively it may be that M_2 or M_3 is capable of forming a more favourable transition state; i.e. a four-centered one as presented above.

Since experimentally this possibility cannot be differentiated from others mentioned, further theoretical approaches to this problem have to be awaited.

In 1986, in work by yet another group [16], dimethyl ether was added to the reaction mixture to investigate the intrinsic reactivity of magnesium surfaces toward bromomethane.

Chemisorption and subsequent decomposition of bromomethane on a Mg(0001) single crystal surface under ultra high vacuum conditions were studied using low-energy electron diffraction (LEED), Auger electron spectroscopy (AES), temperature-programmed decomposition (TPD) and high-resolution electron loss spectroscopy (EELS).

The conclusion from this work is that stable surface alkyls are *not* observed, even at temperatures as low as 123 K. Furthermore coadsorbed dimethyl ether does not perturb the reaction pattern. At this temperature cleavage of the carbon-bromine bond is facile and the process is largely insensitive to solvation effects, but such bond-cleavage does *not* lead to the formation of a stable magnesium-carbon bond.

The EELS-data, strongly indicate that methyl moieties are not retained on the surface at 123 K (even at very small bromomethane exposures). This may happen through a facile homolytic cleavage of the Mg–C bond as has been observed for the surface-moderated decomposition of trimethylaluminium [17].

Obviously the barrier to the Grignard formation is greater than the barrier for the desorption of the multilayer. The question suggested therefore becomes one of whether outer-sphere electron-transfer processes, mediated by such layers, are important in the formation of organomagnesium reagents. The authors' reply is:

We simply don't know. Future studies may help establish the importance of such notions in this and related systems.

Adsorption of Halides on Aluminium. Very recently the adsorption of iodoalkanes [18] as well as of 1,3-dihalo-alkanes [19] on Al(100) and Al(111) surfaces was studied.

Alkyl ligands are stable on the surface at temperatures as high as 450 K after which thermal decomposition takes place via β-hydride elimination.

It is not yet clear to what extent these reports will contribute to a better understanding of the formation reaction of organometallics in bulk solvents.

4.2.4 Conclusion

Modern techniques and newly developed instrumentation allow a closer observation of surface phenomena on metals on which organo halides as well as other reactants are (co-)adsorbed.

In view of the elucidation of the mechanism of the Barbier reaction it is to be hoped that also coadsorption on the metal surface of carbonyl compounds together with organo halides will be studied.

4.3 Mechanism of the Grignard Reagent Formation Reaction

4.3.1 Introduction

The mechanism of the Grignard reagent formation reaction has been the subject of intensive study for quite some time.

The notion, that radicals played an important role in heterogeneous reactions of this type was mentioned at the very beginning of the development of organometallic chemistry: Frankland discovered the formation of organozinc compounds when searching for, what in those days were named 'organic radicals' (see chap. 1).

In his book *Le Magnésium en Chimie Organique*, Courtot surveyed the progress in Grignard Chemistry, a quarter of a century after the discovery of the mixed organomagnesium compounds.

In discussing some sidereactions that occur during the formation of the Grignard reagent, no mention is made of any involvement of radical species [20].

Inspired by earlier research, and theoretical considerations in which monovalent magnesium species were suggested as intermediate species in the formation reaction of the Grignard reagent, Gomberg and Bachmann [21], one year later, presented their ideas about the mechanism of this reaction.

Monovalent magnesium iodide as well as radical-type alkyl groups played a crucial role in the mechanism these authors proposed.

Kharasch and Reinmuth, in 1954, in what has been, until the present day, a standard text on Grignard chemistry [22], propose radical intermediates in the Grignard reagent formation reaction, albeit calling their proposals 'admittedly speculative'.

They suggested that these radicals originated from certain 'points of insaturation' on the magnesium surface, caused by defects in the crystalline structure of the metal: ·Mg(Mg)bulk.

Beside radicals, such as R·, 'groups' such as Mg-X and Mg-R are formed on these 'points of insaturation' and react as given in the following equations:

$$Mg\bullet + R{-}X \longrightarrow (Mg{-}X) + R\bullet$$
$$Mg\bullet + R\bullet \longrightarrow (R{-}Mg)$$
$$(R{-}Mg)(Mg{-}X) + R{-}X \longrightarrow 2RMgX$$

The authors had little experimental support for their speculations, although their reaction schemes accounted for most of the observed sideproducts.

4.3.2 Studies on Radical Formation

A few years after the publication of Kharasch and Reinmuth's text, more systematic studies on radical formation during the reaction of an organic halide with magnesium were started.

Use of a Radical Scavenger: 1958. The reaction of 1-iodobutane with magnesium in 1-methyl-1-phenylethane was investigated [23]; this solvent also functions as a radical scavenger, and from the products so observed conclusions were drawn regarding the formation of radicals in various reaction steps.

Neophyl Radicals: 1962. A few years later, the reaction of 'neophyl chloride' (I) with magnesium was studied [24] in diethyl ether.

Analysis of the products found in the reaction mixture led the authors to the conclusion that their results *per exclusionem* speak for the occurrence of 'neophyl radicals' and their rearrangement.

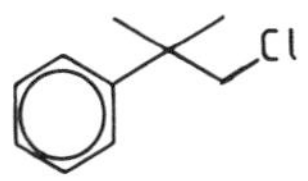

Cl

neophyl chloride (I)

Reaction of Bromobenzene with Magnesium; Detailed Reaction Scheme: 1963. Accurate analysis of the reaction mixture of bromobenzene and magnesium in diethyl ether [25], by yet another research group led to the rejection of an ionic mechanism which had been suggested earlier [26]; a rather detailed reaction scheme for the Grignard reagent formation was presented which – in the authors' view – was close to the mechanism proposed by Kharasch and Reinmuth.

In the first step of the reaction (*Initiation*) an electron is transferred from the metal to the adsorbed halide molecule thus generating a phenyl radical and a bromine anion:

Initiation:

$$PhBr + Mg: \longrightarrow Ph^{\bullet}\ldots Br^{\ominus} + Mg^{\bullet\oplus}$$

The organic radical, depending on its activity and lifetime, can migrate along the metal surface and be further reduced.

The reaction continues in the following manner:

Propagation:

$$PhBr + Mg^{\bullet\oplus} \longrightarrow Ph^{\bullet}\ldots Br^{\ominus} + Mg^{2+}$$

$$\longrightarrow Ph^{\bullet} + Br^{\ominus} + Mg^{2+}$$

$$Ph^{\bullet} + Mg: \longrightarrow Ph^{\ominus} + Mg^{\bullet\oplus}$$

$$Ph^{\ominus} + Mg^{2+} + Br^{\ominus} \longrightarrow PhMgBr$$

It is of importance to note that the authors attempted to describe the initial steps of the reaction mechanism. That is, the electron transfer from the metal to the halide, and its consequences for later reaction processes.

This crucial step was to be studied in more detail by other research groups (see previous paragraphs on the interactions of co-deposited metals and organo halides on pp. 140–142 of Sect. 4.2).

Chiral Organometallics; 1964-today. A more detailed scheme that accounted for the formation of a variety of products observed during the Grignard reagent formation reaction, was presented [27], [31–37] in the 1960s by Walborsky and coworkers.

Use was made of a very useful probe, i.e. (+)-(*S*) 1-bromo-1-methyl-2,2-diphenyl-cyclopropane (II).

II

Earlier attempts to produce chiral organometallic reagents leading to products with measurable optical activity (see e.g. [28–30] had failed, until, in 1961 [31] chiral products were obtained from organometallic derivatives of II.

Although the optical purity of several products, derived from this halide through reactions with magnesium, was not higher than 12–14% it was proved [32] that the Grignard reagent, once it was formed, was optically stable, so that racemization had occurred at some stage preceding the Grignard reagent formation step (see also [33, 34]).

To account for all the observed phenomena the following scheme was proposed [35]:

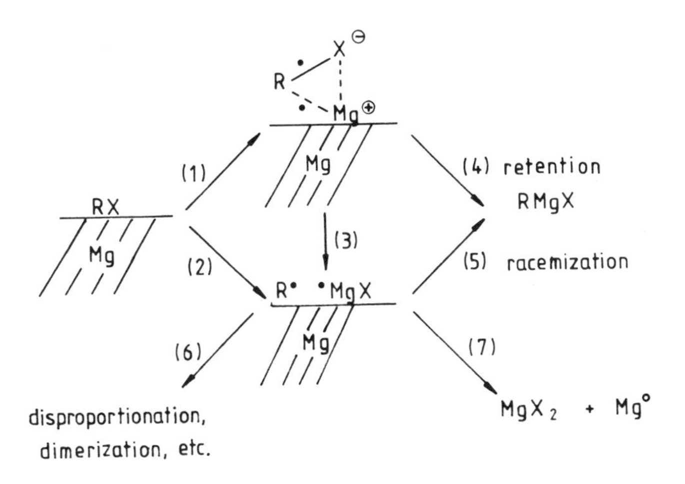

The processes pictured take place at the magnesium-solution interface. Interaction of the cyclopropyl halide and the magnesium in reaction step (1) involves electron transfer from the metal into the antibonding orbital of the carbon-halogen bond. This results in the formation of a radical-anion which is closely associated with a univalent magnesium cation.

When such a '*tight radical pair*' leads to the formation of the Grignard reagent (step (4)) complete retention of configuration of the cyclopropyl entity results.

Alternatively, from step (2) as well as from step (3) a different radical pair can be formed, which was named a '*loose radical pair*'.

It is during this loose radical pair stage, prior to the formation of the Grignard reagent (step 5) that racemization can take place.

Sideproducts will be formed when radicals escape capture by the magnesious halide as represented by step (6) and also in step (7).

In a more recent publication [36] on electron transfer from metal surfaces, further confirmation of this scheme was given, based on the results of reactions of (+)-(*S*)-bromomethyledene-4-methylcyclohexane (III) with magnesium.

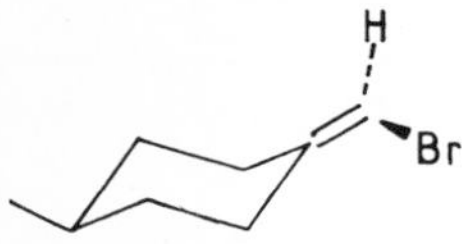

H

Br

III

The stereoselectivity of the reaction, in which an sp^2 hybridised carbon atom is involved, is considerably higher than in the reaction with the cyclopropyl bromide (in which the carbon atom is $sp^{2.28}$ hybridised).

The optical purity of the carboxylic acid, formed on carbonation of the reaction mixture, was 42%.

It should also be mentioned that when lithium was used as the metal in these reactions essentially the same observations were reported [36, 37].

However, in general, the retention of configuration around the carbon atom carrying the halide was higher.

To conclude Walborsky's contribution to the elucidation of the surface reactions in the formation process of organometallic compounds, it remains to be seen to what extent a better understanding can be obtained of the difference between '*tight*' and '*loose*' radical pairs.

CIDNP-Studies: Radical Pairs; 1972–1980. Direct evidence for the presence of radicals in Grignard reagent formation reaction mixtures was obtained with the use of NMR-spectroscopy by Blomberg, Bickelhaupt and coworkers [38] in the beginning of the 1970s.

Chemically Induced Dynamic Nuclear Polarization spectra (CIDNP) were observed when magnesium was reacted with iodoethane in di-*n*-butyl ether or with 2-methyl-1-bromopropane in THF.

Unlike ESR spectroscopy, CIDNP phenomena reflect properties of products derived from radicals and they can be observed over a period which is much longer than the radical lifetime.

From the radical pair theory of CIDNP [39, 40] it can be established that CIDNP phenomena in reaction products are proof for the occurrence of radical pair intermediates.

In a more detailed, subsequent article [41], it was reported that it was the radical pair (A)

$\overline{R\bullet\ \bullet R}$

(A)

which was involved in the formation of the polarized products and not radical pairs such as (B) or (C):

$\overline{R\bullet\bullet MgX}$ or $\overline{RMg\bullet\bullet X}$

(B) (C)

The following scheme represents the complex of reactions:

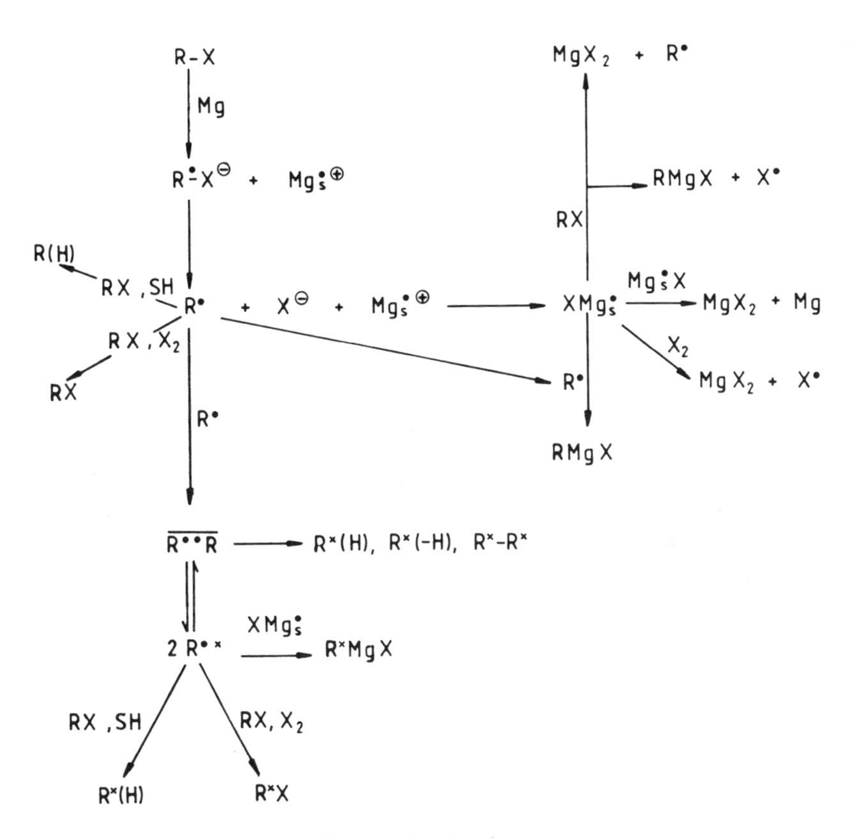

A reasonable explanation for the reaction proceeding via radical intermediates of type (A) and not (B) or (C) may be that XMgs· lacks the required degree of freedom to take part in singlet-triplet mixing in such a radical pair. This is why the subscript 's' in the radical XMgs· was employed to describe it as a surface-bound 'non-free' radical.

From the solvent effects, found in the CIDNP spectra [42, 43] it seems likely that XMgs· is drawn out of the metal surface by the ether before or during the formation of the carbon-magnesium bond.

This is an important conclusion in view of the mechanistic aspects of the Barbier reaction in which the polar organic reagent could take over the role of the solvent.

Radical reaction products, derived from the reagents, which are often found in Barbier reactions, may result from interaction of the radical XMgs· with the carbonyl group of the substrate.

Once XMgs· is drawn out of the surface of the metal, reaction to form RMgX should then be so fast that the conditions for radical pair polarization cannot be fulfilled.

From the scheme it can be seen that the radical R·, which has diffused away from the site of formation, can take part in several reactions.

Among these is the formation of the radical pair $\overline{R \cdot \cdot R}$, responsible for all the polarization observed in the different products.

Continued studies [44–46] of polarization patterns in different reaction products gave increasing support for the scheme, presented above.

Although the Landé splitting factor, *g*-factor, of ·MgBr is unknown it could at least be made evident (e.g. by using mixtures of halides in reaction with magnesium) that radical pairs such as $\overline{R \cdot \cdot MgX}$ cannot be the source of polarization in the products [46].

Studies with Bulky Reagents; 1977–1988. New contributions to the elucidation of the mechanism of the Barbier reaction soon followed the previous ones. Molle, Bauer and Dubois and coworkers discovered [47] that when the reaction mixture of 1-bromoadamantane (IV) (Ad-Br) and magnesium was not stirred the desired Grignard reagent could be obtained in yields varying from 25% to 60% depending on the solvent used (lowest in THF, highest in diethyl ether and in di-*n*-butyl ether).

Br

(IV) 1-Bromoadamantane

Under the usual Grignard reaction conditions only diadamantane (Ad-Ad) and adamantane (Ad-H), from radical attack on the solvent, were formed.

Based on a kinetic analysis of the reaction mixture these authors proposed the following reaction scheme to account for the formation of different products:

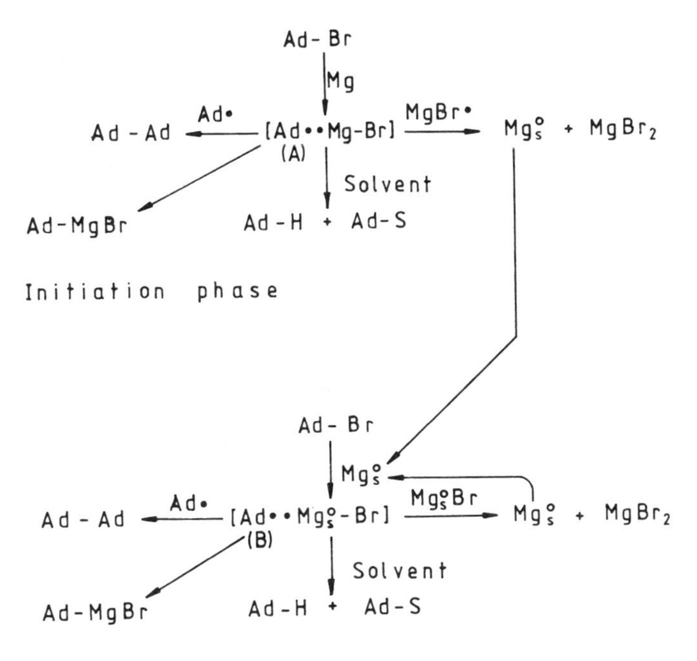

In the first step a complex (A) is formed which mainly leads to undesired sideproducts such as Ad-Ad and solvent-reaction products.

Unreacted monovalent magnesium bromide, ·MgBr, disproportionates to form magnesium bromide and finely divided magnesium Mgs°, which covers the surface. The barely soluble Ad-Ad forms a compact deposit which partially protects the zero-valent magnesium from attack by Ad-Br (see also [48] in which X-ray photoelectron spectroscopic analysis of the black layer on the metal is presented). It is this highly reactive magnesium that gradually forms complex (B) from which the Grignard reagent results.

In a third paper [49] the firm conclusion was drawn that:

.. *the formation of organomagnesium compounds might be a typical surface reaction, whereas side reactions would occur in the medium, and competition between these two reaction pathways would depend on the degree of adsorption of the transient species at the metal surface; the degree of adsorption would be dependent on the electrostatic interactions between the transient species and the metal, after the SET. If this is so, the degree of adsorption, and therefore the reaction rates and competition between the pathways, should be influenced by the nature of the halogenated derivatives of group R, and of the solvent and by the area of the metal surface.*

As was the case with the research of Walborsky et al., Molle, Bauer, Dubois and coworkers found that the formation reaction of organolithium compounds showed great similarities to the formation reaction of organomagnesium compounds [50–56]. Their work in this field covered a period of ten years and one very remarkable result – i.e. the inability of the bulky organo-lithium compounds they investigated to react with sterically hindered ketones under certain conditions – will be discussed in Sect. 4.4.

More Recent Studies: Percentage of Radicals Formed and their Mobility in the Metal Surface. A significant contribution to a better understanding of the Grignard reagent formation reaction came, more recently, from Whitesides' group; among others a series of kinetic studies was published [57–62].

In one specific report [62] the reaction of bromocyclopentane with magnesium was studied in the presence of the radical trap 2,2,6,6-tetramethylpiperidine nitroxyl (TMPO) and *tert*-butyl alcohol which functions as a proton donor.

The following scheme illustrates the reactions involved:

TMPO (slow)

$$\text{R-Br} \xrightarrow{\text{Mg}} [\text{R}\bullet] \xrightarrow{\text{Mg}} \text{RMgBr}$$

$[\text{R}\bullet] \xrightarrow{\text{TMPO}} \text{TMPOR}$; $\text{RMgBr} \xrightarrow{t\text{-BuOH (fast)}} \text{RH}$

More than 80% TMPOR was formed, and this, the authors stated

establishes that at least this fraction of the starting RBr is converted into radicals (n-hexyl bromide → 70% in a similar experiment)

These observations, combined with those made during CIDNP experiments as reported above, led the authors to conclude:

are all compatible with free (as opposed to 'surface-bound') alkyl radicals (the differences between a hypothetical 'surface-bound' and a free radical in solution have always been so vague as to frustrate experimental distinction. In the experiments reported here, any surface-bound radicals would have to have properties very similar to those of free radicals).

Whitesides, together with Garst and Deutch more recently [63] reported the results of studies in which ratios of several rearranged (R'·) and unrearranged (R·) radicals in Grignard reagent formation reactions were measured:

$$RX + Mg \longrightarrow R^{\bullet}$$

$$R^{\bullet} \xrightarrow{k} R'^{\bullet}$$

$$R^{\bullet} \longrightarrow RMgX$$

$$R'^{\bullet} \longrightarrow R'MgX$$

They stated that their results;

conform to predictions based on a model in which the intermediate alkyl radicals diffuse freely in solution. This "D Model" is closely related to the detailed mechanism favored by Bickelhaupt and co-workers.

One intriguing question is why the alkyl radicals, if not surface bound, do not undergo more reaction with the solvent during the Grignard reagent formation.

The simple answer lies in the differences in rate constants:

It is difficult for even a freely diffusing radical to escape reaction at a sufficiently reactive surface when it is generated near that surface.

Radical probes that were capable of cyclization were recently used to study [64] the extent to which radicals leave and return to the surface of magnesium to form a Grignard reagent. In THF 6-bromo-1-hexene reacted with magnesium to form approximately 90% of the expected Grignard reagent and 5% of the cyclized product. Protium-containing hydrocarbons were also formed resulting from radical reaction with the solvent:

Br + Mg —THF / 2) $D_3O^{\oplus}$→ D (90%) + (1%) + D (5%) + (0.8%)

When the same reaction was executed in THF in the presence of 10 equivalents of dicyclohexylphosphine, which donates a protium atom much more readily than THF, the following yields were observed.

D (71%) + (24%) + D (0.01%) (4.6%)

The authors therefore concluded that:

.. of the Grignard reagent formed in this reaction, a minimum of 25% is a result of radicals that diffuse into the solution and then return to the magnesium surface.

At the beginning of the 1990s, regarding the different views on the mechanism of the Grignard reagent formation reaction, the question was in short formulated as follows [65]:

Does Grignard reagent formation proceed entirely via free radicals, or are other intermediates such as radical anions involved?

The D Model (see also a later publication by Garst et al. [66]) allows "all radicals to leave the surface and diffuse freely in solution all times".

On the basis of further studies of reactions of bromocyclopropanes [65, 67] Walborsky et al. came to the conclusion [65] that the mechanism

involves ... a tight radical anion-radical cation pair as well as a loose radical pair all adsorbed on the surface of the magnesium with only a small percentage of the radicals leaving the surface.

and that their results, in no way allows that all radicals leave the surface and diffuse freely in solution.

Conclusion. There is a general agreement on the existence of radicals on the metal surface in the Grignard reagent formation reaction. A large part of the halide is converted to a radical anion followed by heterolytic cleavage of the carbon-halogen bond.

However, discussion still continues as to the 'freedom' of these 'free' radicals. That is, to the distance they can diffuse away from the metal surface and still retain a high probability to return (or at least to an MgX· entity) in order to form the RMgX molecule.

Perhaps the quotation (p. 150) from Lawrence and Whitesides' paper [62] is an appropriate conclusion:

... the differences between a hypothetical 'surface-bound' and a free radical in solution have always been so vague as to frustrate experimental distinction.

4.4 Mechanism of the Barbier Reaction

4.4.1 Introduction

The number of publications dedicated to the mechanism of the reaction in which an organic halide reacts with a metal in the presence of a third reactant is very limited indeed.

In this Section an attempt will be made to present an explanation of the phenomena observed during some of these studies and to give an account of the (type of) products formed.

Use will be made of information from the two previous Sections,
a) the interaction of organic halides with metals and
b) the mechanism of the Grignard reagent formation reaction.

Intermediacy of Organometallics. It is illustrative to quote some of the final words of a rather recent paper [68] on the magnesium-induced cyclization of 2-(3-iodopropyl)-cycloalkanones, an intramolecular Barbier reaction (see also Sect. 2.4.2, p. 46 in chap. 2):

The Barbier reaction is generally formulated in terms of formation of an organometallic intermediate, followed by addition of this species to the carbonyl compound in the usual manner. However there is little mechanistic evidence to bolster this point of view, and in one instance of an analogous reaction using lithium instead of magnesium, evidence against such a process has been reported.

In this last sentence the authors referred to reports [52, 53] on a one-step synthesis including 1-bromoadamantane and adamantanone:

Br + O + Li → 2) $H_3O^{\oplus}$ HO
40%

If a two-step procedure is employed virtually no tertiary alcohol is formed.

However, earlier studies [69] on the kinetics of the *One-step alternative to the Grignard reaction* of bromobenzene, benzaldehyde and lithium in THF had led to the conclusion that

the data are consistent with a rate-determining step involving intermediate formation of phenyl-lithium by reaction of bromobenzene with lithium.

In that same period systematic studies [70] of the Barbier-type reaction of allylic bromides, benzaldehyde and zinc in ethanol as the solvent (see Sect. 3.4.4, p. 111) also had led to the conclusion that the organometallic reagent was actually involved as an intermediate in these reactions.

4.4.2 Radical Formation

From what has been reported in Sect. 4.3 about the Grignard reagent formation reaction it is not unreasonable to suggest that great similarities will be found in the mechanisms of Barbier-type reactions with either magnesium, or lithium or zinc or, for that matter, with several other metals.

Two Pathways. Single electron transfer (SET) from the metal to either the organic halide or the substrate (often, but not always, an aldehyde or a ketone), or perhaps even to both the reactants, is most likely to be the initial step.

These two steps are to be found in the scheme presented by Molle and Bauer [53] for the Li-Barbier reaction:

Reactions (1) and (2) involve single electron transfers from the metal to either the ketone or the halide. In the latter case this leads to the formation of what the authors named *the precursors of the organolithium compound or transitory species on the metal surface.*

By extending this study to other cage-structured radicals, it was demonstrated that *the radical pathway* (4) is in competition with *the organometallic pathway* (3), and that one of the factors that regulate this competition is the stability of the radical.

On the ground of the experimental evidence published by Pearce, Richards and Scilly in 1970 [71], 1972 [72] and 1974 [73] (see Sect. 3.3), Molle and Bauer rejected the possibility of product formation via the reaction of the ketyl radical with the halide (see however also Garst and Smith in 1976 [74]):

$$RX + Ar_2\dot{C}{-}O^{\ominus}M^{\oplus} \longrightarrow R^{\bullet} \xrightarrow{Ar_2\dot{C}-O^{\ominus}M^{\oplus}} Ar_2C(O^{\ominus}M^{\oplus})R$$

However, there is still need for clarification of such terms as *transitory species on the metal surface or precursors of organometallic compounds.*

Is this merely a more modern – but still unsatisfactory – term for what in the past had been named "nascent" organometallics (see e.g. Bryce-Smith and Wakefield's [75] earlier proposal of such "nascent" organometallic compounds)?

From Sect. 4.3 it has become evident that the degree of freedom of the radicals on the metal surface is still

so vague as to frustrate experimental distinction [62].

Furthermore theoretical questions arise with regard to what has been reported in Sect. 4.2 about "clusters" of metals to be involved in the first SET steps on the metal surface. Does the transfer from a single electron take place simultaneously to the halides as well as to the ketone through such clusters?

Evidence for Radicals from Ketones and Aldehydes. In the reaction of Grignard reagents such as the bulky neopentylmagnesium bromide with benzophenone considerable amounts of benzopinacol can be formed [76]:

BrMg + =O —— 2) $H_2O^{\oplus}$ → OH + HO— —OH

20%

See [77] and [78] on the formation of pinacols from reactions of organometallic reagents with (substituted) benzophenones. Indeed benzopinacol was found [79] among the products of almost all the Barbier reactions listed in Tables 2.18 through 2.21 in Sect. 2.4.4. Radical formation on the surface of the metal is most likely the origin of such sideproducts.

Also the reports of the Barbier reaction with benzaldehyde in Section 2.4 [79] show the formation of products which can only be rationalized when radicals are assumed to be intermediates.

Products of the following (possible) sidereactions were found:

$$R\text{-}X + Mg \longrightarrow R^{\bullet} + MgX^{\bullet} \qquad (1)$$

$$R''\text{-}CHO + R^{\bullet} \longrightarrow R''\text{-}CO^{\bullet} + R\text{-}H \qquad (2)$$

$$R''\text{-}CO^{\bullet} + R^{\bullet} \longrightarrow R''\text{-}CO\text{-}R \qquad (3)$$

$$R\text{-}X + Mg + R''\text{-}CO\text{-}R \xrightarrow[2)\ H_3O^{\oplus}]{} R''C(OH)R_2 \qquad (4)$$

Dimerization of the acyl radical leads to the formation of a 1,2-diketone:

$$2R''\text{-}CO^{\bullet} \longrightarrow R''\text{-}CO\text{-}CO\text{-}R'' \quad (5)$$

On rapid addition of the reactants (bromobenzene and benzaldehyde) causing a high density of unreacted species on the metal surface (see Table 2.24; the author described that reaction as performed in a "careless" way) only a low yield of the expected secondary alcohol was obtained; benzophenone was isolated in a 40% yield together with 20% of the tertiary alcohol (triphenylmethanol) derived from this ketone (reactions (3) and (4)).

In a parallel reaction, performed in a "careless way", using deuterated benzaldehyde, C_6H_5CDO, 10% deuterated benzene was found among the products.

Another complication is the hydrogen abstraction from the aldehyde by the monovalent magnesium species, MgX·. This leads to the formation of a magnesium hydride derivative:

$$R''CHO + MgX^{\bullet} \longrightarrow R''CO^{\bullet} + HMgX \quad (6)$$

which reduces aldehydes to primary alcohols [80]:

$$R''CHO + 2HMgX \xrightarrow[2)\ H_3O^{\oplus}]{} R''CH_2OH \quad (7)$$

Indeed the formation of more than 20% benzyl alcohol was reported in the reaction of bromo- or iodoethane with magnesium and benzaldehyde (Table 2.23) and 30% in the 'uncontrolled' reaction of bromobenzene with magnesium and benzaldehyde (Table 2.24).

A reaction which might interfere in Barbier reactions with aldehydes, is the direct reaction of magnesium with 2 equivalents of the aldehyde.

$$2RCHO + Mg \longrightarrow RCO\text{-}COR + MgH_2 \quad (8)$$

Although this reaction hasn't been observed under Barbier conditions it was reported [81, 82] to occur at room temperature under nitrogen without a solvent.

Substitution in Aromatic Rings. Another proof for the existence of radicals as intermediates in the Barbier reactions is given by the detection of products which result from aromatic substitution reactions: see Table 2.7 and Tables 2.18 through 2.21.

In 1942 Fuson et al. [83] observed that in the reaction of phenylmagnesium bromide with 2,4,6-trimethylbenzophenone a ring-substitution product, 2,4,6-trimethyl-2′-phenyl-benzophenone, was formed, apparently as the result of a 1,4-addition reaction.

Both 1,4- as well as 1,6-addition reaction products were observed almost thirty years later [84] in reactions of benzophenone and *t*-butylmagnesium chloride in diethyl ether:

=O + ClMg

2) $H_3O^{\oplus}$

OH H

OH H

HO

Such products are easily oxidized by air to the substituted benzophenones.

Kinetic studies [85, 86] have led to the conclusion that the rate determining step in this type of reaction is a single electron transfer from the organomagnesium species to benzophenone under the formation of the ketyl radical (see also further references in [85, 86]).

CIDNP in Barbier Reactions. A direct proof for the occurrence of radicals in Barbier reactions would be given by the observation of CIDNP effects in the NMR-spectra of the reaction mixture (see Sect. 4.3, pp. 146–148 [38–46]).

Figure 4.1 shows a typical example of CIDNP observed during the Grignard reaction of iodoethane with magnesium in benzene and one molar equivalent of THF [79].

The so-called 'multiplet effect' in the resonances of the protons in an α-position to the magnesium $\delta \approx -0.5$ ppm) is clearly visible. The 'net' emission effect at $\delta \approx 1$ ppm is probably due to ethane and butane. This effect is frequently observed though as yet not fully understood.

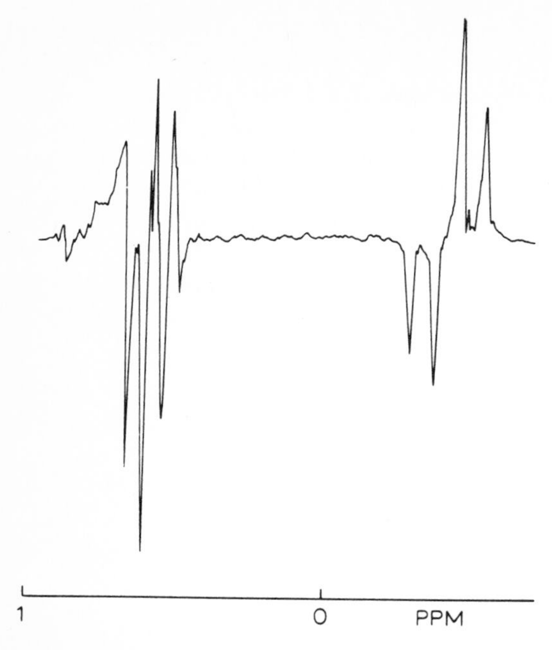

Fig. 4.1. Observed CIDNP during the Grignard reaction of iodoethane with magnesium in benzene and one molar equivalent of THF [79]

From the Barbier reaction mixture containing bromo- or iodoethane, magnesium and the bulky 2,4-dimethyl-3-pentanone in a THF/benzene solvent mixture, only weak CIDNP signals were observed in the region of about − 0.5 ppm.

No CIDNP effect could be detected when benzophenone was applied in this one-step reaction. This may imply that:

a) no in situ Grignard compound is formed because of the rapid reaction of benzophenone with either R· or MgX· or
b) the steady-state concentration of "free" ethylmagnesium halides is too low to be observed. This might be due to a more rapid reaction with benzophenone as compared with 2,4-dimethyl-3-pentanone.

CIDNP was clearly observed however in the signals of the Grignard reagent as well as of the hydrocarbons in the Barbier reaction of benzyl benzoate with iodoethane (Fig. 4.2).

For a better understanding of the origin of a 'multiplet effect' in the 2–3 ppm region (not shown in Fig. 4.2) this reaction was performed with iodomethane and benzene as the solvent. Surprisingly no additional THF was needed to keep the reaction mixture clear. The NMR-spectrum is shown in Fig. 4.3

Figure 4.3 shows two multiplet effects: a quartet at $\delta \approx 2.5$ ppm and a triplet at $\delta \approx 1.1$ ppm.

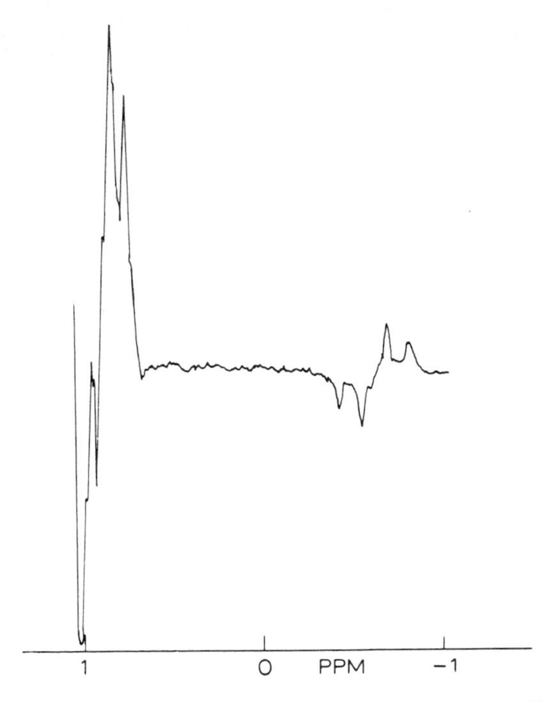

Fig. 4.2. Observed CIDNP during the Barbier reaction of iodoethane, magnesium and a threefold excess of benzyl benzoate in THF/benzene [79]

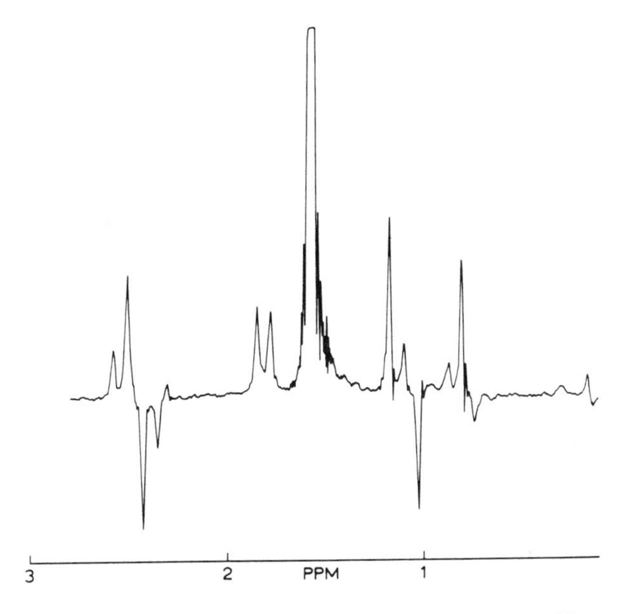

Fig. 4.3. Observed CIDNP in the reaction of iodomethane with magensium and an excess of benzyl benzoate in benzene [79].

The large peak in the middle ($\delta \approx 1.4$ ppm) is due to unreacted iodomethane.

The NMR spectrum taken when the reaction was finished, indicates few absorptions in the regions where the CINDP phenomena had been observed. Gaschromatographic analysis of the reaction mixture indicated that ethylbenzene had been formed in low yields (less than 10%) and it is this product that must be held responsible for the observed CIDNP phenomena.

Its formation from benzyl- and methyl-radicals as presented in the following equation, leads to the polarizations observed:

$$CH_3^{\bullet} + C_6H_5CH_2^{\bullet} \longrightarrow \overline{CH_3^{\bullet}\bullet CH_2C_6H_5} \longrightarrow CH_3CH_2C_6H_5$$

The benzyl radical probably originates from the reaction of the ester with the activated (by etching) magnesium;

Mg; $O^{\ominus}$; $+ Mg^{\bullet\oplus}$; O; OMg$^{\bullet}$; +; O; O; Mg

Although such a reaction with magnesium is not known Katzenellenbogen and Lenox have reported that lithium and geranyl mesitoate react in THF to form geranyllithium in good yields [87].

The phenomena observed in the above-mentioned reaction present the first example of the trapping of the radical from the alkyl halide other than by the same alkyl radical or by MgX·. Thus an independent proof for the occurrence of alkyl radicals during the Grignard and the Barbier reaction has been obtained.

Effects of the 'Activation of the Metals'. In the last decade several reports on the mechanism of the Barbier reaction have appeared in which *not* magnesium but lithium and zinc were the metals applied.

It was mentioned in chap. 3 that very early in the history of zinc in organic reactions it was found that the metal needed some sort of activation; this was done either by preliminary treatment of the metal with acid or iodine, or by amalgamation with other metals such as sodium or copper.

No theoretical foundation for this practice is known but probably cleaning the surface and/or creating some 'points of insaturation' (see Sect. 4.3, p. 143) was sufficient to promote the reaction.

Some quotations from chap. 3 may be presented here to demonstrate that this idea of an active surface comes up regularly in mechanistic considerations of Barbier-type reactions:

a. In the preparation of perfluoroalkane carboxylic acids via a Zn-Barbier reaction [88] (Sect. 3.4.3, p. 102) it had become evident that the reactivity of the organometallic intermediate is increased on the metallic surface since the preformed perfluoralkylzinc halide does not react with carbon dioxide.
b. From the results of Zn-Barbier reactions in which Zn/graphite was used [89] (Sect. 3.4.3, p. 124) the conclusion was drawn that the organozinc compound (which was suggested to be formed as an intermediate) was activated by the graphite support and that the reaction with the substrate (a nitrile in this case) occurs at the surface of the carbonaceous matrix rather than in the solution.
c. For the *electrochemical Grignard-type allylation of carbonyl compounds* [90] (Sect. 3.5, p. 128) use was made of aluminium powder and Sn[II] chloride. The results suggested that zero-valent tin alone is not enough to give a satisfactory reaction but that the zero-valent tin on a metallic aluminium surface is responsible for the efficient metallation with allyl chloride.

It can only be speculated what mechanisms apply on the metal liquid interface but recent publications in the field of Li-Barbier reactions may shed more light on the functioning of the substrate in Barbier-type reactions.

Stereochemical Studies: *1987*. In a stereochemical study of ultrasound-supported Li-Barbier reactions of *S*(+) 2-halooctane and cyclohexanone, Luche and co-workers obtained an optically active alcohol in each of the reactions performed [91]. Its enantiomeric excess and absolute configuration strongly suggested the existence of two reactive intermediates following different pathways to form the condensation alcohol. The following scheme was given to account for the results:

R-C-H X Li• 1 A1 2 A2 3 B desorption

R = n-C_6H_{13}

R(+) product *R*(+) product *S*(-) product

The 'tight radical pair' A1 (see Sect. 4.3, p. 145) reacts with the ketyl radical anion with preferential inversion at the chiral center. In case of weak adsorption on the metal surface A1 should racemize through the easily desorbed species A2. Under conditions which favour both electron transfer and step 2 (higher temperatures) formation of B is accelerated, especially for the more reactive species (X = Cl). Retention of configuration will thus progressively become more important (or inversion less important).

Effects of Sonication. In a subsequent study [92] on the accelerating effect of ultrasonic waves on the Barbier reaction of benzaldehyde, 1-bromo-heptane and lithium, Luche and co-workers could not state that cavitation was the only important phenomenon. No information had been obtained on the effects of non-cavitational shock waves or the frequency. The authors came to the conclusion that the work published had to be considered as an approach for a better knowledge of the interaction of ultrasound with a heterogeneous system, but the problem in its generality could not be considered as fully understood.

Semiempirical MO Calculations. A more detailed picture of the radical ion mechanism of the Barbier reaction was given by Luche and co-workers [93] using semiempirical MO calculations. A theoretical support was provided for the observed configuration inversion at the halide carbon center as found in the reaction of $S(+)$ 2-halooctane mentioned above.

The following scheme was proposed:

R-X + R'C(O)R'' —SET 1→ (A) $[R-X]^{\bullet\ominus}$ + R'C(O)R'' ; (B) R-X + $[R'C(O)R'']^{\bullet\ominus}$ → $[R'C(O)R''\text{-----}R\text{-}X]^{\bullet\ominus}$ complex 1

complex 1 → $[R'R''C(O)\text{-}R\text{--}X]^{\bullet\ominus}$ complex 2 —SET 2→ $R'R''C(R)O^{\ominus}$ + $X^{\ominus}$

Whether Path A or Path B will be followed can be predicted from the electron affinities of the reactants, since the radical anion actually formed will be that derived from the species with the higher electron affinity. Some values for such affinities were presented.

With saturated carbonyl compounds participating in Barbier reactions the initial step should be the formation of the halide radical anion (Path A).

4.4.3 Conclusion

Radical formation is evident in Barbier reactions as can be concluded from radical reaction products.

Single electron transfer from the metal can take place both to the organo halide as to the carbonyl compound.

Recent work on the mechanism of the Barbier reaction indicates that the formation of an (intermediate?) organometallic species is not a necessary step in this one-step process.

On the other hand, the observation of CIDNP in an intermediate organometallic compound is direct proof for its existence though its steady-state concentration could be small. Furthermore, the third reactant in this particular one-step reaction was an ester and not a ketone or an aldehyde as usual.

Semiempirical calculations have led to the conclusion that electron transfer from the metal to (one of) the reactants is dependent on the electron affinity of the compounds involved and thus leads to different reactions and hence to different mechanisms.

4.5 References

1a. Horn E, Polanyi M and Style DWG (1934) Trans Farad Soc 30: 189
1b. Warhurst E (1951) Quart Rev (London) 5: 44
2. Skell PhS and Doerr RG (1967) J Amer Chem Soc 89: 4688
3. Skell PhS, Wescott Jr LD, Golstein J-P and Engel RR (1965) J Amer Chem Soc 87: 2829
4. Timms PL, J Chem Comm 1968: 1525
5. Skell PhS and Girard JE (1972) J Amer Chem Soc 94: 5518
6. Klabunde KJ, Key MS and Low JYF (1972) J Amer Chem Soc 94: 999
7. Klabunde KJ, Low JYF and Key MS (1972–1973) J Fluor Chem 2: 207
8. Klabunde KJ (1975) Acc Chem Res 8: 393
9. Wescott Jr LD, Williford C, Parks F, Dowling M, Sublett S and Klabunde KJ (1976) J Amer Chem Soc 98: 7852
10. e.g. Ternay Jr AL (1976) Contemporary Organic Chemistry, Saunders WB Company, 2nd edn, p 680
11. Ault BS (1980) J Amer Chem Soc 102: 3480
12. Tanaka Y, Davis SC and Klabunde KJ (1982) J Amer Chem Soc 104: 1013
13. Imizu Y and Klabunde KJ (1984) Inorg Chem 23: 3602
14. Jasien PG and Dykstra CE (1983) J Amer Chem Soc 105: 2089
15. Klabunde KJ and Whetten A (1986) J Amer Chem Soc 108: 6529
16. Nuzzo RG and Dubois LH (1986) J Amer Chem Soc 108: 2881
17. Squire DW, Dulcey CS and Lin MC (1985) Chem Phys Lett 116: 525
18. Bent BE, Nuzzo RG, Zegarski BR and Dubois LH (1991) J Amer Chem Soc 113: 1137
19. Bent BE, Nuzzo RG, Zegarski BR and Dubois LH (1991) J Amer Chem Soc 113: 1143
20. Courtot Ch (1926) Le Magnésium en Chimie Organique, Lorraine-Rigot et Cie, Nancy
21. Gomberg M and Bachmann WE (1927) J Amer Chem Soc 49: 236

22. Kharasch MS and Reinmuth O (1954) Grignard Reactions of Non-Metallic Substances. Prentice Hall, New York. p. 61
23. Bryce-Smith D and Cox GF, J Chem Soc 1958: 1050
24. Rüchardt C and Trautwein H (1962) Chem Ber 95: 1197
25. Anteunis M and van Schoote J (1963) Bull Soc Chim Belg 72: 787
26. Prévost Ch, Gaudemar M, Miginiac L, Bardone-Gaudemar F and Andrac M, Bull Soc Chim France 1959: 769
27. Walborsky HM and Young AE (1964) J Amer Chem Soc 86: 3288
28. Schwartz AM and Johnson JR (1931) J Amer Chem Soc 53: 1063
29. Porter CW (1935) J Amer Chem Soc 57: 1436
30. Tarbell DS and Weiss M (1939) J Amer Chem Soc 61: 1203
31. Walborsky HM and Young AE (1961) J Amer Chem Soc 83: 2595 (1961)
32. Walborsky HM, Impastato FJ and Young AE (1964) J Amer Chem Soc 86: 3283 and references cited therein.
33. Walborsky HM, Chen C-J and Webb JL, Tetrahedron Lett 1964: 3551
34. Walborsky HM and Chen C-J (1971) J Amer Chem Soc 93: 671
35. Walborsky HM and Aronoff MS (1973) J Organometal Chem 51: 31
36. Walborsky HM and Banks RB (1980) Bull Soc Chim Belg 89: 849
37. Walborsky HM and Aronoff MJ (1965) J Organometal Chem 4: 418
38. Bodewitz HWHJ, Blomberg C and Bickelhaupt F, Tetrahedron Lett 1972: 281
39. Closs GL (1969) J Amer Chem Soc 91: 4552
40. Kaptein R and Oosterhoff LJ (1969) Chem Phys Lett 4: 214
41. Bodewitz HWHJ, Blomberg C and Bickelhaupt F (1973) Tetrahedron 27: 719
42. Bodewitz HWHJ, Blomberg C and Bickelhaupt F (1975) Tetrahedron 31: 1053
43. Bodewitz HWHJ, Blomberg C and Bickelhaupt F, Tetrahedron Lett 1975: 2003
44. Schaart BJ, Bodewitz HWHJ, Blomberg C and Bickelhaupt F (1976) J Amer Chem Soc 98: 3712
45. Bodewitz HWHJ, Schaart BJ, van der Niet JD, Blomberg C, Bickelhaupt F and den Hollander JA (1978) Tetrahedron 34: 2523
46. Schaart BJ, Blomberg C, Akkerman OS and Bickelhaupt F (1980) Canad J Chem 58: 932
47. Dubois JE, Bauer P, Molle G and Daza J (1977) Compt Rend Sér C 284: 145
48. Dubois JE, Molle G, Tourillon G and Bauer P (1979) Tetrahedron Lett 20: 5069
49. Molle G, Bauer P and Dubois JE (1982) J Org Chem 47: 4120
50. Molle G, Dubois JE and Bauer P (1978) Synth Comm 8: 39
51. Molle G, Dubois JE and Bauer P (1978) Tetrahedron Lett 19: 3177
52. Bauer P and Molle G (1978) Tetrahedron Lett. 19: 4853
53. Molle G and Bauer P (1982) J Amer Chem Soc 104: 3481
54. Molle G, Bauer P and Dubois JE (1983) J Org Chem 48: 2975
55. Dubois JE, Bauer P and Kaddani B (1985) Tetrahedron Lett 26: 57
56. Dubois JE, Bauer P and Briand S (1988) Tetrahedron Lett 29: 3935
57. Rogers HR, Hill CL, Fujiwara Y, Rogers RJ, Mitchell HL and Whitesides GM (1980) J Amer Chem Soc 102: 217
58. Rogers HR, Deutch J and Whitesides GM (1980) J Amer Chem Soc 102: 226
59. Rogers HR, Rogers RJ, Mitchell HL and Whitesides GM (1980) J Amer Chem Soc 102: 231
60. Barber JJ and Whitesides GM (1980) J Amer Chem Soc 102: 239
61. Root KS, Deutch J and Whitesides GM (1981) J Amer Chem Soc 103: 5475
62. Lawrence LM and Whitesides GM (1980) J Amer Chem Soc 102: 2493
63. Garst JF, Deutch JE and Whitesides GM (1986) J Amer Chem Soc 108: 2490
64. Ashby EC and Oswald J (1988) J Org Chem 53: 6068
65. Walborsky HM and Zimmermann C (1992) J Amer Chem Soc 114: 4996
66. Garst JF, Ungvary F, Batlaw R and Lawrence EL (1991) J Amer Chem Soc 113: 5392
67a. Walborsky HM and Rachon J (1989) J Amer Chem Soc 111: 1896
67b. Rachin J and Walborsky HM (1989) Tetrahedron Lett 30: 7345
67c. Walborsky HM (1990) Acc Chem Res 23: 286
68. Crandall JK and Magaha HS (1982) J Org Chem 47: 5368
69. Cameron GG and Milton AJS, J Chem Soc Perkin Trans. II, 1976: 378
70. Killinger TA, Boughton NA, Runge TA and Wolinsky J (1977) J Organometal Chem 124: 131
71. Pearce PJ, Richards DH and Scilly NF, J Chem Soc D 1970: 1160
72. Pearce PJ, Richards DH and Scilly NF, J Chem Soc Perkin Trans I 1972: 1655
73. Pearce PJ, Richards DH and Scilly NF (1974) British Patent 1335, 1975, 140. Chem Abstr 80: 70306

74. Garst JF and Smith CD (1976) J Amer Chem Soc 98: 1520
75. Bryce-Smith D and Wakefield BJ, Tetrahedron Lett 1964: 3295
76. Blomberg C and Mosher HS (1968) J Organometall Chem 13: 519
77. Ashby EC, Buhler JD, Lopp IG, Wieseman TL, Bowers Jr JC and Laemmle JT (1976) J Amer Chem Soc 98: 6561
78. Okubo M (1977) Bull Chem Soc Japan 50: 2379
79. Hartog FA (1978) Thesis Free University Amsterdam. Available on request.
80. Ashby EC and Goel AB (1977) J Amer Chem Soc 99: 310
81. Givelet MHE (1968) Compt Rend Sér C 278: 881
82. Givelet MHE (1969) French Patent 1541058, (1968); Chem Abstr 70: 25265y
83. Fuson RC, Armstrong MD and Speck SB (1942) J Org Chem 7: 297
84. Holm T and Crossland I (1971) Acta Chem Scand 25: 56
85. Ashby EC, Laemmle JT and Neumann HM (1974) Acc Chem Res 7: 272
86. Ashby EC and Laemmle JT (1975) Chem Rev 75: 521
87. Katzenellengbogen JA and Lenox RS (1973) J Org Chem 38: 326
88. Blancou H, Morreau P and Commeyras A, J Chem Soc Chem Comm 1976: 885
89. Boldrini GP, Savoia D, Tagliavini A, Trombini C and Umani-Ronchi A (1983) J Org Chem 48: 4108
90. Uneyama K, Matsuda H and Torii S (1984) Tetrahedron Lett 25: 6017
91. de Souza-Barbosa JC, Luche J-L and Pétrier C (1987) Tetrahedron Lett 28: 2013
92. de Souza-Barboza JC, Pétrier C and Luche J-L (1988) J Org Chem 53: 1212
93. Moyano A, Pericàs MA, Riera A and Luche J-L (1990) Tetrahedron Lett 31: 7619

5 Experimental Procedures for Barbier Reactions

5.1 Introduction

In general, not only experimental skill and experience are required to carry out a planned organic synthesis. Some precautions have to be taken and some conditions have to be fulfilled–the so-called 'tricks' of the specialist–in order to obtain the best possible results.

Among the earliest reports of such specialist's tricks to promote a smooth procedure in which an organo halide and a metal were involved, was Pébal's publication on the preparation of diethylzinc in 1861 [1] (see Sect. 1.2.2, p. 3). For the reaction of iodoethane with zinc in diethyl ether he used zinc that had first been 'etched' ('*angeätz*') with sulfuric acid, washed and carefully dried.

Furthermore the diethyl ether was first made 'completely' anhydrous with 'anhydrous phosphoric acid'.

That the trick did not always work as indicated in the report may probably be concluded from the fact that, one year later, Rieth and Beilstein [2] recommended the use of a 4:1 zinc/sodium alloy for the same reaction.

Many of the readers of this monograph, if not all of them may recognize the starting problems in routine organometallic reactions. Such problems evidently were encountered from the very beginning of organometallic chemistry.

More than hundred years after Pébal's and Reith and Beilstein's reports on starting problems of an organometallic reagent formation reaction, Dreyfuss [3] noted that the start of Barbier reactions with allylic halides

was found to be critical to the success of this method.

First a small amount of the allyl halide had to be added to the magnesium in diethyl ether before beginning the addition of allyl halide and functional addend.

An unawareness of this technique,

the author continued,

probably accounts for the preference for the two-step procedure by workers such as Henze [4] *and Bacon and Farmer* [5].

However, not only starting problems were encountered in Barbier reactions and other organometallic reactions. Problems could also arise from, e.g. an 'uncontrolled' addition of the reagents to the metal magnesium in a Barbier reaction (Table 2.24, Sect. 2.4.4, p. 66; see also the discussion of these reactions in Sect.4.4.2, p. 156). Such reaction conditions caused the formation of unexpected and often also undesired sideproducts.

This practical chapter in a monograph on the Barbier reaction will deal with some experimental procedures and such special conditions – mainly with regard to the activation of the metal – required for the optimal execution of Barbier and Barbier-type reactions.

5.2 Starting Heterogeneous Reactions with Metals

5.2.1 Introduction

Not only metallic zinc causes problems in its reactions with organo halides; also magnesium and even more reactive metals such as lithium have their specific difficulties.

Tissier and Grignard, in one of the earliest publications on Grignard chemistry [6], found that for the reaction of magnesium with bromobenzene, iodobenzene and analogous organo halides, the metal usually had to be activated by the addition of a 'crystal' of iodine.

Bayer and Villiger, shortly after Grignard and Tissier, reported [7] that reactions of bromo- and iododimethylaminobenzene with magnesium could not be brought about, even after boiling the reaction mixture for several days.

Sachs and Ehrlich, in the same year [8], managed to get such reactions going by first 'etching' the metal through preliminary reaction with bromoethane in diethyl ether after which the main portion of the ethereal solution was removed and the 'activated' magnesium was used further.

The start of reactions of organo halides with metallic lithium also often gave great problems:

In 1959, three different research groups reported unsatisfactory results from reactions with 'pure' lithium for the preparation of organolithium compounds:

a. Several attempts to prepare *tert*-butyllithium from 'sodium-free' lithium failed. Good results were obtained [9] when 1–2% sodium had been added to the metal in the melt, in preparing lithium sand.
b. For the preparation of lithium derivatives of 4-dialkylamino-1-bromobenzene the most successful grade of lithium was the one containing approximately 0.6% sodium [10]. Experiments with newer low-sodium grades (0.02 – 0.005%) were less successful.
c. A convenient method of preparation of *n*-butyllithium was reported when lithium, containing 0.8% sodium was used [11].

It was stated that:

the amount of sodium in the lithium metal appreciably affects the yield of n-butyllithium from n-butyl halides and lithium metal, though the reason for this effect is not clear. The sodium has to be intimately mixed with the lithium as the addition of pieces of sodium or sodium sand to the 'low sodium' lithium does not change the yield.

Lithium with 2% sodium was used for the preparation of 4-dimethylaminophenyllithium and *tert*-butyllithium [12] as well as for the preparation of vinyllithium [13].

In the following, attention will be focussed on the metals magnesium, zinc and lithium respectively in their reactions with organohalides. Unfortunately only in a limited number of the reports on the reactions of those metals, are Barbier reactions involved.

Other metals than those three will be mentioned briefly where such information is relevant.

One preliminary general remark has to be made, apart from the pretreatment of the metals in one-step processes.

This is in regard to:

5.2.2 Anhydrous Reaction Conditions

It is generally accepted in organometallic chemistry that strict anhydrous reaction conditions are a *conditio sine qua non* for a smooth reaction and that therefore the glass vessel, reflux-condenser, stirrer, dropping funnel, metal chips, etc. will have been dried for several hours in an electric oven before use.

The organic reagents as well as the solvent will have to be freshly distilled, the latter preferably from a solution containing an organometallic reagent, lithium aluminium hydride or any other agent that removes water and 'acid-hydrogens'.

Such rigorous conditions were already prescribed in Courtot's book on twenty five years of Grignard chemistry as published in 1926 [14].

It will be understood in the whole of this chapter that strictly anhydrous reaction conditions will be applied in Barbier-type reactions unless otherwise stated. [Water was used as part of the solvent, e.g., in Zn-Barbier reactions (See Sect 3.4, pp. 112–114).]

Such conditions for this type of reaction have been extensively mentioned and discussed in standard textbooks for organometallic chemistry; it is therefore sufficient here to refer to these books for further guidance and instruction [15].

5.3 Reactions with Magnesium

5.3.1 Introduction

Sachs and Ehrlich (vide supra p. 167) used bromoethane to facilitate the reaction with magnesium before the organo halide was applied of which an organomagnesium derivative was desired.

More modern is the use of 1,2-dibromoethane as an 'entrainment reagent' with the disadvantage that is consumes an extra amount of magnesium;

furthermore the magnesium bromide which is formed is less soluble in diethyl ether and may cause a two-layer system.

The use of iodine as an 'activator' leads to the formation of magnesium iodide which forms a reductive couple with metallic magnesium, Mg/MgI_2, which may lead to pinacol formation when a ketone is added later to the reaction mixture.

Such problems may be found to be even more serious in Barbier-type reactions where the ketone is present in the reaction mixture from the very beginning.

For 'difficult' Grignard reagent formation reactions Bayer, in 1905, recommended [16] the use of magnesium with enhanced reactivity, prepared by heating the metal with half its weight (ca 0.05 mole equivalents) of iodine.

Apart from activation with iodine, which still finds wide application in preparative Grignard and Grignard-type chemistry, a great variety of possibilities to activate magnesium has been proposed, used, rejected and again reintroduced and a whole chapter of this monograph could be filled with this information. However, for a summary, see the standard reference books for Grignard and organozinc chemistry [15].

It is probably true that there is a consensus among specialists in the field of organomagnesium chemistry about one quality required for magnesium when used in Grignard and Barbier reactions.

The magnesium applied should be of the highest possible purity in order to obtain the best results.

Contrary to the case for lithium (vide supra), impurities in magnesium hamper its reactivity.

Freshly sublimed magnesium [17]–such crystals, usually about 3 to 10 mm in length, are as lustrous as platinum–in general leads to the most satisfactory and reproducible results. It has to be fully admitted here, however, that sublimation of magnesium is no *sinecure* under ordinary laboratory conditions.

Since the number of systematic publications in the field of Barbier chemistry is rather limited also the number of reports on the use of activated magnesium in Barbier reactions is very small indeed. Nevertheless, since, in general, also with this type of reaction the start of the procedure may be the critical step, there is good reason to make mention of such techniques that have been helpful in two-step Grignard-type reactions in order to obtain the best possible results with one-step processes.

5.3.2 Surface of Metallic Magnesium

It seems obvious that a smooth Barbier reaction is promoted by a clean metal surface; mention has been made (vide supra) of the advantage of shiny, freshly sublimed magnesium crystals. In Sect. 4.2 a discussion was presented of reports, in the beginning of the 1970s, on Grignard reagent formation reactions at low temperatures by co-condensation of magnesium vapour and haloalkanes on a cold surface.

A supremely clean and active metal surface seems to be formed during the cutting of chips from a rod of metallic magnesium at the point of a cutting tool, as reported in a patent as early as 1945 [18].

The apparatus described facilitates a Grignard type reaction by continually cutting chips from the metal in the presence of an appropriate solvent together with the organohalide which reacts on the clean nascent chip surface.

Continued research [19] in this field, which was named 'mechanical activation', demonstrated its advantages. The Grignard reaction starts extremely smoothly and

The usual precautions required to maintain the apparatus and reagents absolutely anhydrous are unnecessary when the more easily prepared Grignard reagents are produced by this procedure.

The newly invented technique proved to be quite valuable, not only in initiating a wide variety of Grignard reagent formation reactions but–as the author put it–it also

illustrates the manner in which a Barbier type of reaction in which the two steps in the Grignard synthesis are performed simultaneously, may be carried out by mechanical activation

Unfortunately no example of such a Barbier reaction was given.

Although the author was probably correct in his final statement that

there is every indication...that mechanical activation will enable certain organometallic reactions to compete commercially with the current methods of synthesis...

there is a clear underestimation of the technical difficulties of cutting a magnesium rod under common laboratory conditions in the statement that

Only additional research and experience with this relatively new technique will establish the position it is to assume in the field of organic synthesis.

A much simpler procedure [20] for the activation of magnesium involves the one day (or longer) room temperature mechanical stirring of magnesium turnings in a nitrogen atmosphere; addition of 4-dimethylaminobromobenzene in THF leads to the formation of the expected Grignard reagent

without the aid of an initiation agent such as iodine or an entrainment reagent such as a haloalkane

Very recently, in a similar procedure [21], magnesium metal turnings were vigorously stirred mechanically under an inert atmosphere in a Schlenk tube with the help of a Teflon-coated stirrer bar. Both nitrogen and argon gave the same results which demonstrated that surface magnesium nitrides do not play a part in the activation process. Sufficient diethyl ether (freshly distilled off sodium benzophenone ketyl) was run in to cover the magnesium after which the organo halide was introduced.

This method had proved very effective in synthesis of several allyl- and benzylmagnesium halides free from coupling products; neither sonication (vide infra) nor the use of different grades of magnesium powder had given satisfactory results.

Although the authors did mention sonication as

the preferred technique for effecting the Barbier variation, in which a carbonyl compound and an organohalide are introduced concomitantly and the Grignard reagent is then intercepted as fast as it is formed

unfortunately no efforts were reported to try out this 'activated' magnesium in Barbier reactions.

Almost twenty years earlier mechanically activated magnesium had been used [22] in an intramolecular Barbier reaction of diethyl 5-bromo-3-bromomethylpentylphosphonate in THF, leading to 1-phosphabicyclo [2.2.1] heptane 1-oxide in 11% yield:

Br
O
Br P(-OEt)$_2$ + Mg —THF→ P O

11%

Using a completely different technique, a highly reactive metal surface is formed on evaporation of the metal and condensation on a cold surface. Results of investigations on the reactivity of such metals were reported in Sect. 4.2. Co-condensation of ^{1}S, ground-state, magnesium with water [23] gave more than 90% hydrogen after warming up

$$^1S\text{-}Mg + H_2O \longrightarrow H_2$$

and codeposition of organo halides at −110 °C yielded solvent-free Grignard reagents (see Sect. 4.2 p. 138 for results of such reactions).

In a U.S. Patent a new method was claimed [24] for the preparation of organometallic compounds by using an electric arc to comminute metal, thus producing the metal in highly reactive, finely divided form. Among the examples given was the preparation of highly reactive magnesium which was applied for the preparation of phenylmagnesium chloride in a hydrocarbon fraction, together with a small amount of pyridine, as the solvent.

A few years later an apparatus was described [25] in which a highly reactive magnesium slurry was prepared by evaporation of magnesium at 5×10^{-3} bar from an alumina crucible in a simple rotating-solution reactor (Fig. 5.1).

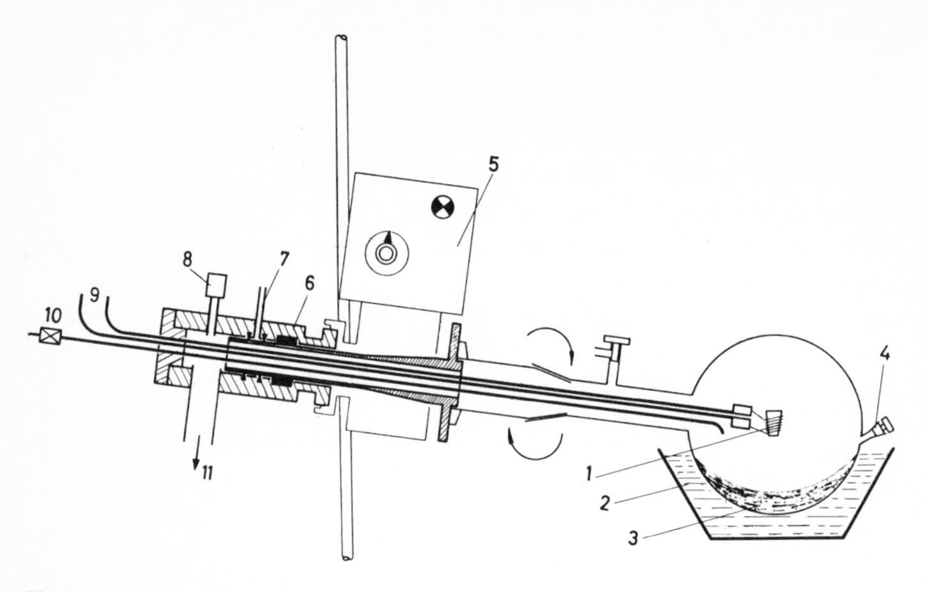

Fig 5.1. Rotating solution reactor as used for the preparation of Mg-slurry for Grignard reagent formation at low temperatures [24] *1*: alumina crucible. *2*: cooling bath. *3*: reaction flask with solvent. *4*: septum cap. *5*: rotatory drive. *6*: cylindrical bearings. *7*: lubricated Viton Symerings. *8*: pressure guage. *9*: water-cooled electrodes. *10*: substrate inlet tube with needle valve. *11*: to high-vaccum pump

After condensation of the magnesium-atoms into THF at $-110\,^{\circ}C$ the black Mg-slurry was used, at $-75\,^{\circ}C$, for reaction with bromo-cyclopropyl-methane followed by condensation with carbon dioxide into the reaction mixture at $-110\,^{\circ}C$.

On hydrolysis cyclopropylmethanoic acid and 4-pentenoic acid were isolated:

Br — Mg (slurry), THF/ -75 °C → MgBr → MgBr

+ CO_2, 2) $H_3O^{\oplus}$ + CO_2, 2) $H_3O^{\oplus}$

OH, O   OH, O

92 : 8

Overall yield 78%

Substantial ring-opening occurs when this Grignard reaction is carried out at higher temperatures [26–28].

In later publications [29, 30] from the same laboratory smooth metallation of allylic halides with this highly active, precondensed magnesium in THF was reported. A version of the apparatus, shown in Fig. 5.1, was reported to be commercially available.

5.3.3 New Developments in Activation of Magnesium

In the last two decades several new developments in the field of 'activated metals', also including magnesium, have been introduced.

Varying claims have been made as to which of these techniques would be the 'easier' or the 'more practical one'; unfortunately, all together, the number of investigations in which such 'activated forms of magnesium' were used in the field of Barbier-type reactions is very limited indeed.

Sonication. Sonication implies the use of ultrasound in chemistry and related disciplines. Only very recently has the first volume appeared of a research annual, reporting advances in sonochemistry [31].

Sonication was successfully introduced in organometallic chemistry in 1980 [32] when it was applied to Li-Barbier reactions (see Sect. 3.3.7, pp. 90–94); it was found that ultrasonic irradiation allows this reaction to be performed in wet technical grade THF. This compares well with the findings that freshly cut magnesium chips also react with haloalkanes in wet diethyl ether (see p. 170).

The efficiency of sonochemical removal of adsorbed water and alcohol from magnesium surfaces was demonstrated [33] in the reaction of 2-bromopentane with this metal in diethyl ether: when this solvent was half-saturated with water, an induction period of only 6–8 minutes was required before the reaction started. Without sonication it took 1–3 hours refluxing in the same solvent to get this reaction going.

The authors reported that ultrasound was used routinely in their general laboratory work to initiate reluctant Grignard reactions.

It was claimed [34] that three hours of ultrasonic irradiation was required in 'Grignard-type reactions' of vinyl bromides with magnesium in THF while trifluoroacetaldehyde was bubbled into the reaction mixture. Unfortunately, no comparison was made of results, obtained with other techniques in this specific type of reactions.

Ultrasound was used [35] for a one-pot (though not one-step) preparation of the etherate of triethylaluminium in 80% yield: a mixture of bromoethane, aluminium and magnesium powders was irradiated with ultrasound and ethylaluminium sesquibromide was formed. As soon as diethyl ether was introduced into the reaction medium, ethylmagnesium bromide was formed which reacted in situ with the aluminium compound to yield triethylaluminium etherate.

$$3\,EtBr + 2\,Al \xrightarrow{\cdot)))} \underset{\mathrm{I}}{Et_3Al_2Br_3}$$

$$3\,EtBr + 3\,Mg \xrightarrow{\cdot)))} 3\,EtMgBr + \mathrm{I} \xrightarrow{Et_2O} Et_3Al.OEt_2$$

Ultrasound dramatically accelerated the formation of triorganylboranes [36] in a Barbier reaction of an organo halide, magnesium and BF_3. etherate. Yields with various halides were

1-bromopropane	100%	1-bromobutane	100%
2-bromopentane	96%	bromocyclohexane	99%
bromobenzene	97%	1-bromonaphthalene	93%
benzylchloride	99%	allyl chloride	90%
1-iodoheptane	90%		

It was also reported, however, that in spite of the use of ultrasound the yields of tri-*n*-butyl borane were only 30% with 1-chlorobutane as the starting halide and 82% with the iodide.

Also for the clean synthesis of reactive allylic of benzylic organomagnesium chlorides [21] neither sonication (nor the use of different grades of magnesium powder, for that matter) provided a satisfactory solution and activated magnesium was required (see p. 171).

Sonication was effectively used [37] for synthesis of *N*-substituted benzamides in one-step reactions of bromobenzene or 2-methylbromobenzene, *tert*-butyl isocyanate and a metal in diethyl ether or THF. With magnesium yields were almost quantitative. However, chlorobenzene proved to be unreactive under the same reaction conditions.

In conclusion sonication has proved to be a powerful new development in the field of Barbier chemistry. (see e.g. [31]).

In view of unexpected and unpredictable variations in its effectiveness to initiate, and promote one-step reactions (vide supra) more systematic research in this field seems to be required in order to justify a statement, as quoted on p. 71 [21], that

sonication is the preferred technique for effecting the Barbier variation.

Rieke Magnesium. 'Rieke metals' are highly reactive metal powders prepared by reduction of the corresponding halide by potassium in a suitable solvent. The first publication in this field by Rieke and co-workers appeared in 1972 [38]. For reviews see [39].

Rieke magnesium, Mg*, allows the formation of Grignard reagents at low temperatures. (*S*)-(+) 1-bromo-1-methyl-2,2-diphenylcyclopropane e.g., on reaction with Mg* at − 65 °C, yields a chiral Grignard reagent that is 33–43% optically pure [40].

Such 'active magnesium' reacts at room temperature with butadiene [41] to yield an organomagnesium compound which, after the addition of silicon tetrachloride, yields 5-silaspiro[4,4]-nona-2, 7-diene:

$$2\ \text{butadiene} + 2\ Mg^{*} \longrightarrow 2\ \text{(magnesacyclopentene)} \xrightarrow{SiCl_4} \text{5-silaspiro[4,4]nona-2,7-diene}$$

When the reaction was carried out with 'ordinary' magnesium in hexamethylphosphoric triamide heating was required for several days.

However, Barbier reactions with isoprene and dialkoxydichlorosilanes [42], gave excellent yields of 2, 7-dimethyl-5-silaspiro-[4, 4]nona-2, 7-dienes when just 'ordinary' magnesium was used.

Recently reactions of dienemagnesium reagents with 1, *n*-dibromoalkanes and bromocyanoalkanes were published [43] which led to the formation of spiroalkanes and -ketones in an intramolecular Barbier reaction through an intermediately formed halo- or cyano-substituted Grignard reagent:

$$\text{diene} + Mg^{*} \longrightarrow \text{dienemagnesium} \xrightarrow{Br(CH_2)_{n+2}Br} \text{MgBr / Br intermediate} \longrightarrow \text{spiroalkane}$$

Initial attempts to perform such reactions with 'ordinary magnesium' were not successful.

Magnesium Anthracene. Magnesium powder is activated by catalytic amounts of anthracene in THF. Although the reaction of anthracene and magnesium was discovered relatively long ago [44] only in the 1980s did the chemistry of magnesium anthracene systems start to develop; for a review see [45].

Allylmagnesium chloride can be generated at $-78\,^{\circ}C$ from allyl chloride and magnesium powder in the presence of 2 mol % magnesium anthracene [46].

Substituted allyl Grignard reagents were synthesized in high yields from the corresponding chlorides and Mg-slurry, obtained by evaporation, activated by equilibration with its anthracene adduct in THF [30]. The authors compared this magnesium with 'Rieke-magnesium' and found, on careful experimentation, that also the Rieke-method gave satisfactory results.

Their paper ends with the following remarks:

We may thus conclude that, .. the magnesium evaporation technique offers the advantage of choice of solvent. The use of 'Rieke–magnesium' requires tedious, stoichiometricallly precise experimentation and is not very attractive for working on a very small or large scale.

So far, the use of highly activated Mg-slurry in the presence of anthracene appealed to these authors as the most practicable, widely usable approach to allylic Grignard reagents and their reaction products.

Magnesium-Graphite. Finally, magnesium-graphite, Mg/Gr, is to be introduced in this section as a newly developed type of activated magnesium. It is obtained by the reduction of MgI_2 with potassium graphite, C_8K, in diethyl ether or THF [47] (for a review on graphite-metal compounds see [48]):

$$MgI_2 + 2\ C_8K \longrightarrow C_{16}Mg(KI)_2$$

Reaction of Mg/Gr with organo halides allows the preparation of Grignard reagents in excellent yields (95–100%) even at temperatures as low as – 78 °C (with an equilibrium between the intercalated (40–60%) and soluble Grignard compounds).

Even fluorobenzene forms the corresponding Grignard compound at a temperature of – 20 °C. Although in the review article, no mention is made of Barbier reactions carried out with Mg/Gr, it was mentioned (from unpublished work) that

the high reactivity of the magnesium-graphite reagent even allows Reformatsky-type reactions .. to yield the corresponding Mg ester enolate.

A two-step procedure, however, was given as an illustration of such a 'Magnesio-Reformatsky' reaction.

The general applicability of Mg/Gr for this type of reaction (applied in the carbohydrate field) seemed to be restricted but, the authors claimed that

Rieke Mg was shown to be insufficiently reactive for this purpose

5.3.4 General Conclusion Regarding Magnesium-Activation

In general it can be stated that modern developments in the preparation of active magnesium have been extremely helpful to overcome starting problems of reactions of organo halides with this metal.

The preparation of Grignard reagents at low temperatures, even the preparation of phenylmagnesium fluoride at – 20 °C, demonstrates the versatility of such developments.

Unfortunately, application of these new techniques and these new, active, forms of magnesium has not yet been studied to a large extent in one-step processes; the examples published so far, however, had promising results albeit that they lack sufficient theoretical background for their interpretation.

It is to be expected that future work may shed more light on the relatively unexplored field of Mg-Barbier reactions.

5.4 Reactions with Lithium

5.4.1 Introduction

The use of lithium in Barbier reactions was successfully introduced at the end of the 1960s. A considerable number of such reactions have been published since then (see Sect. 3.3).

In view of what has been said about the reactivity of the lithium towards organohalides in relation to its purity (pp. 167 and 168) it may seem surprising that, for one-step reactions, the quality of the metal doesn't seem to be of crucial significance.

In the Experimental Part of the first important paper on Li–Barbier reactions [49], it was specifically mentioned that

lithium metal was obtained at various levels of purity.

On the other hand, in studies [50] on the optical purity of organolithium compounds derived from (+)-(*S*) 1-bromo-1-methyl-2, 2-diphenylcyclopropane (see Sect. 4.3.2, pp. 144–146) it was demonstrated that the nature of the lithium surface, as well as the purity of the metal did influence the results.

5.4.2 Activation of Lithium Metal

Variations in the properties of the surface of metallic lithium are limited to the particle size of Li-sand, prepared by vigorously stirring molten lithium in an inert solvent, followed by cooling. Particle sizes varying from 25 μ to 150 μ were used in the above-mentioned studies on the optical purity of organolithium compounds [50a, 50b].

One very important recent development in the use of lithium in Barbier-type reactions has drawn great attention though, i.e. the application of ultra-sound.

Sonication. The introduction of sonication (for a recent review on sonication see [31]) in Li-Barbier chemistry [32] has enhanced the synthetic importance of this procedure:

1. The process is speeded up considerably.
2. Wet technical grade solvent could be used at room-temperature, and
3. direct in situ formation of benzylic, allylic and vinylic organolithium compounds were also possible leading to ketone-addition products in good yields.

In a recent study [51] on some fundamental aspects of the sonochemical Barbier-type reaction of benzaldehyde, 1-bromoheptane and lithium it was found (see also Sect. 4.4, p. 132) that the rate of product formation depended strongly on the intensity of the ultrasonic waves and the temperature; for both parameters an optimum was observed. An unusual variation was found for the product formation rate with temperature, which made it clear that the reaction is not mass-transport controlled.

It could not be stated – in conclusion – that cavitation is the only important phenomenon when sonication is applied. Furthermore, no information had been obtained on the effect of noncavitational shock waves or on the frequency.

This work then has to be considered as an approach for a better knowledge of the interaction of ultrasound with a heterogeneous system, but the problem in its generality cannot be considered as fully understood.

Not only has Li-Barbier chemistry increased tremendously in importance in recent years, the additional introduction of sonication in this field has made this procedure of even greater synthetic value (see also [31]).

Addition of an Electron Transfer Agent. In one of the earliest reports on Li-Barbier chemistry [49], a comparison was made of results obtained with lithium and with sodium in one-step processes. It was found that yields with sodium could be improved by the addition of an electron acceptor such as naphthalene or biphenyl in stoichiometric as well as in catalytic amounts.

An advantageous use of an electron acceptor in a Li-Barbier reaction is demonstrated in the following example.

Although Li-Barbier reactions had proved to be very effective to bring about syntheses of sterically crowded alcohols (see Sect. 3.3.6, pp. 86–89) little success was encountered in such a reaction with 2-chloro-2-phenylpropane and adamantone [52]. The desired product, 1-(2-adamantyl)-2-methyl-2-phenyl-1-ethanol, was obtained in 67% yield, however, when 1–3 mol% of 4,4′-di-*tert*-butylbiphenyl (DBB) was added as an electon-transfer agent.

The use of stoichiometric amounts of DBB also gave a good conversion into the product but it proved impossible to separate these two compounds.

It may be of interest to report here that lithium was used containing ca. 3.3% sodium!

Conclusion Regarding Lithium-Activation. The great advantages of the use of lithium in Barbier-type reactions, as demonstrated in Sect. 3.3, have grown even more in importance by the use of ultrasound and the addition of an electron-transfer agent.

More research is required to finally decide whether magnesium should be replaced by lithium in most of the one-step processes it is still used for nowadays (in one example e.g., the one-step synthesis of *N-tert*-butyl benzamide, as mentioned on p. 174 [37], magnesium was the metal of preference over lithium, sodium or potassium).

5.5 Reactions with Zinc

5.5.1 Introduction

In Sect. 3.4 it was demonstrated that the metal zinc has made a glorious come-back to the field of organic synthesis after it had been pushed aside by magnesium in the beginning of this century.

It had remained the favourite metal for Reformatsky reactions though, and was utilized on and off for Grignard-type syntheses – preferably to be named 'Saytzeff reactions' (see Sect. 1.2.4). In modern organic chemistry zinc has come back as the preferred metal in many different types of less traditional reactions, not least because of new developments in activating techniques to overcome its sluggishness in the initiation of such reactions.

This Section will cover these new developments as well as the older, more traditional, ones. As was the case with magnesium and lithium in the previous Sections, it should be mentioned that also not all the examples of zinc-activation presented here have been applied in Zn-Barbier reactions. Several of them come from other organic synthetic procedures such as the Reformatsky or the Simmons-Smith reaction.

5.5.2 Chemical Activation of Zinc

Alloys with Metals. In the beginning of this chapter reference was made to Pébal's [1] advice, in 1861, to activate zinc by 'etching' with sulfuric acid, for its reaction with iodoethane.

In later periods by far the most widely used method for the activation of zinc was alloying it with copper. In one of the standard textbooks for organometallic chemistry [53] the preparation was described of not less than five different Zn/Cu couples. Furthermore three different Zn/Pb couples were mentioned in the same reference, indicating the large number of problems encountered with this metal.

Surprisingly Zn/Cu couples were also used for Zn-Barbier reactions when modern techniques of activation were used to have the reaction starting: e.g. a 50% Zn/Cu couple had proved its superiority over other types of activated zinc [54] in the following reaction:

$$\mathrm{Ph_2P(O)CH{=}CH_2} + \mathrm{RX} \xrightarrow[\mathrm{EtOH/H_2O}]{\text{1:1 Zn/Cu; ·)))}} \mathrm{Ph_2P(O)CH_2CH_2R}$$

In a review of the Simmons–Smith reaction [55] it is stated that the copper plays no other role than activating the zinc surface for reaction, and that in ether solvents such as 1,2-dimethoxyethane zinc metal alone is effective in the cyclopropane ring formation [56].

Another metal, used in alloys with zinc was silver; a Zn/Ag couple was successfully used in a modification of the Simmons–Smith reaction [57].

With Iodine. As is the case with magnesium (see Sect. 5.3.1) iodine has been applied [58] in a Barbier-type reaction with propargyl bromide and cyclohexanone. More recently, in a Reformatsky reaction under sonication [59], use was made of iodine-activated zinc. In the absence of iodine the sonicated mixture of zinc, ethyl 2-bromo-ethanoate and acetophenone reacted slowly and gave no addition product: recovery of the ester and the ketone were quantitative.

With Trimethylchlorosilane. Rather recently [60] it was found that activation of zinc with TMClSi augmented the yield of Zn-Barbier reactions with allylic halides considerably. This unexpected activation by the silane was found earlier for Reformatsky reactions with ethyl 2-bromopropanoate [61]; also trimethylchlorosilane – activated lead was applied in Pb – Barbier reactions with crotyl bromide [62] and allyl bromide [63].

With Ammonium Chloride. For the preparation of *o*-quinodimethanes from the corresponding *o*-dibromomethylaromatic compounds [64] use was made of zinc dust, activated by stirring with a saturated NH_4Cl solution, followed by decantation and successive washing with water, ethanol, diethyl ether and DMF.

The application of this type of activated zinc in a Reformatsky reaction was reported one year later [59].

With Titanium [IV] Chloride. A novel type of Reformatsky reaction was reported [65] when ethyl 2-bromoethanoate was reacted with 2-acetoxytetrahydrofuran derivatives and zinc, activated by the addition of 10% $TiCl_4$ in dichloromethane as the solvent. In the more common solvents such as benzene, toluene or THF, yields were significantly lower.

5.5.3 Other Techniques to Activate Zinc

As was the case with magnesium, also highly reactive forms of zinc were made through the 'Rieke method' and through its intercalation compounds with graphite.

Furthermore also sonication was applied to reaction mixtures with metallic zinc.

Rieke Zinc. Highly reactive metal powders are made through the reduction of metal salts, using alkali metals in a suitable solvent [38, 39]. Rieke zinc, Zn*, has been widely used in Simmons–Smith and Reformatsky-type reactions and also its application in a Zn-Barbier reaction [66] is known.

Zinc/Graphite. Zinc/graphite, Zn/Gr, prepared from C_8K and anhydrous $ZnCl_2$ in refluxing THF under argon has been used successfully in many reactions in which usually 'normal' zinc is applied [48].

An even more reactive form of Zn/Gr is obtained when silver is added: Zn/Ag/Gr [48]. It is of interest to note that Rieke zinc was found unsuitable for glycal formation reactions for which Zn/Ag/Gr was applied successfully [67].

Sonication. Ultrasonic radiation was introduced in Zn–Barbier chemistry in 1981 [68], soon after it was applied for the first time in organometallic chemistry [32] for Li-Barbier reactions: iodotrifluoromethane was reacted with 'commercially available zinc' (no mention was made of any activation of the metal) in the presence of an aldehyde or a ketone in DMF or THF to give the expected trifluoromethyl carbinols in reasonable yields (50–70%).

Astounding results – from the viewpoint of an organometallic chemist – were obtained with sonication a few years later, when allylic halides were allowed to react with zinc and aldehydes or ketones in aqueous media [69]. The highest yields were obtained with aqueous ammonium chloride/THF solvent mixtures. Commercially available zinc was used without any further treatment.

Conjugate addition of alkyl groups to α-enones with the aid of zinc and ultrasound in a one-step reaction was reported more recently [70]. Here too, the solvent contained water, together with either ethanol or 1- or 2-propanol. A sound absorption maximum was found for this reaction which indicates a maximum three-dimensional structure organisation of the liquid. That such an optimal organisation parallels a maximum in the reaction yield, most probably indicates a strong solvent participation in a determining step [70b].

A good resumé of the technical details of ultrasound generators to be used in one-step processes is to be found in references [31] and [71].

5.5.4 General Conclusion Regarding Zinc Activation

It is probably true that activation of zinc by means of modern inventions in the 1980s has had the most dramatic results in changes in its reactivity. Organometallic reactions with water as the solvent were never dreamt of before with zinc as the metal. The modern organometallic chemist must come under the impression that a better understanding of the mechanistic aspects of one-step processes is coming nearer once technical problems as the initiation of such reactions is mastered by simple techniques as the use of a sonicator or of Zn* and Zn/Gr.

More systematic studies of these processes are eagerly awaited in order to get such a better understanding.

5.6 References

1. Pébal L (1861) Ann 118: 22
2. Rieth R and Beilstein F (1862) Ann 126: 245
3. Dreyfuss MP (1963) J Org Chem 28: 3269
4. Henze HR Allen BB and Leslie WB (1942) J Org Chem 7: 326
5. Bacon RGR and Farmer EH, J Chem Soc 1937: 1065
6. Tissier L and Grignard V (1901) Compt Rend 132: 1182
7. Bayer A and Villiger V (1903) Ber 36: 2775
8. Ehrlich P and Sachs F (1903) Ber 36: 4296
9. Stiles M and Mayer PR (1959) J Amer Chem Soc 81: 1497; footnote 38b p. 1501
10. Wright JB and Gutsell ES 1959 J Amer Chem Soc 81: 5193; footnote 10 p. 5193
11. Beel JA, Koch WO Tomasi GE Hermanes DE and Fleetwood P (1959) J Org Chem 24: 2036
12. Kamienski CW and Esmay DL (1960) J Org Chem 25: 1807
13. West R and Glaze WH (1961) J Org Chem 26: 2096
14. Courtot Ch (1926) Le Magnésium en Chimie Organique Lorraine-Rigot & Cie. Nancy, p 21
15a. Kharasch MS and Reinmuth O (1954) Grignard Reactions of Non-metallic Substances Prentice-Hall Inc New York
15b. Joffe ST and Nesmeyanov AN (1967) The Organic Compounds of Magnesium, Beryllium, Calcium, Strontium and Barium; North Holland Amsterdam
15c. Nützel K (1973) In: Methoden der Organischen Chemie (Houben–Weyl) Metallorganische Verbindungen, Vol 13/2a: Georg Thieme, Stuttgart
16. Bayer A (1905) Ber 38: 2759
17a. Bodewitz HWHJ, Blomberg C and Bickelhaupt F (1975) Tetradedron 31: 1053
17b. Seetz JWFL, Hartog FA, Böhm HP, Blomberg C, Akkerman OS, and Bickelhaupt F (1982) Tetrahedron Lett 23: 1497
18. Stevens H (1947) British Patent 571, 539, August 29, 1945; Chem Abstr 41: P1696
19. Shaw MC (1948) J Applied Mechanics 15: 37
20. Mendel A (1966) J Organomental Chem 6: 97
21. Baker KV, Brown JM, Hughes N, Skarnulis AJ and Sexton A, (1991) J Org Chem 56: 698
22. Wetzel RB and Kenyon GL (1972) J Amer Chem Soc 94: 9232
23. Skell Ph S and Girard JE (1972) J Amer Chem Soc 94: 5518
24. Lang JL (1975) US Patent 3 880, 743; April 1975; Chem Abstr 82: P97557
25. Kündig EP and Perret C (1981) Helv Chim Acta 64: 2606
26. Smith LI and McKenzie Jr S (1950) J Org Chem 15: 74
27. Roberts JD and Mazur RH (1951) J Amer Chem Soc 87: 2509
28. Patel DJ, Hamilton CL and Roberts JD (1965) J Amer Chem Soc 87: 5144
29. Oppolzer W, Kündig EP, Bishop PM and Perret C (1982) Tatrahedron Lett 23: 3901
30. Oppolzer W and Schneider Ph (1984) Tetrahedron Lett 25: 3305
31. Mason TJ (1990) (Ed) Advances in Sonochemistry Vol 1 JAJ Press, Greenwich
32. Luche J-L and Damiano J-C (1980) J Amer Chem Soc 102: 7926
33. Sprich JD and Lewandos GS (1983) Inorg Chim Acta 76: L241
34. Ishikawa N, Koh MG, Kitazume T and Choi SK (1984) J Fluorine Chem 24: 419
35. Lin Y-T (1986) J Organometal Chem 317: 277
36. Jadhav PK, Bhat KS, Perumal PT and Brown HC (1986) J Org Chem 51: 432
37. Einhorn J and Luche J-L (1986) Tetrahedron Lett 27: 501
38. Rieke RD and Hudnall PM (1972) J Amer Chem Soc 94: 7178
39a. Rieke RD (1977) Acc Chem Res 10: 301
39b. Rieke RD (1989) Science 246: 1260
40. Rachon J and Walborsky HM (1989) Tetrahedron Lett 30: 7345
41. Salomon RG (1974) J Org Chem 39: 3602
42. Terunuma D, Hatta S, Araki T, Ueki T, Okazaki T and Suzuki Y (1977) Bull Chem Soc Japan 50: 1545
43. Xiong H and Rieke RD (1991) Tetrahedron Lett 32: 5269
44. Ramsden HE, (1968) US Patent 3 354 190, 1967; Chem Abstr 68: 114744
45. Bogdanović (1988) Acc Chem Res 21: 261
46. See [43] footnotes 8, 16 and 17
47. Ungurenasu C and Palie M (1977) Synth React Inorg Met-Org Chem 7: 581

48. Csuk R, Glänzer I and Fürstner A (1988) Adv Organomet Chem 28: 85
49. Pearce PJ, Richards DH and Scilly NF, J Chem Soc 1972: 1655
50a. Walborsky HM and Aronoff MJ (1965) J Organometal Chem 4: 418
50b. Walborsky HM and Aronoff MJ (1973) J Organometal Chem 51: 55
50c. Walborsky HM and Banks RB (1980) Bull Soc Chem Belg 89: 849
51. de Souza-Barboza JC, Pétrier C and Luche J-L (1988) J Org Chem 53: 1212
52. Choi H, Pinkerton AA and Fry JL, J Chem Soc Chem Commun 1987: 225
53. Reference 15c pp 570–574
54. Petrusiewicz KM and Zoblocka M (1988) Tetrahedron Lett 29: 937
55a. Simmons HE, Cairns TL, Vladuchick SA and Haines CM (1973) Organic Reactions 20: 1
55b. Reference 6 in [55a.]: Simmons HE and Smith RD (1959) J Amer Chem Soc 81: 4256
56. Richardson DB, Durrett LR, Martin Jr JM, Putnam WE, Slaymaker SC and Dvoretzky I (1965) J Amer Chem Soc 87: 2763
57. Denis JM, Girard C and Conia JM, Synthesis 1972: 549
58. Zeile K and Meyer H (1942) Ber 75: 356
59. Han B-H and Boudjouk Ph (1982) J Org Chem 47: 5030
60. Picotin G and Miginiac Ph (1987) Tetradedron Lett 28: 4551
61. Gawronsky JC (1984) Tetrahedron Lett 25: 2605
62. Tanaka H, Hamatani T, Yamashita S and Torii S, Chem Lett 1986: 1461
63. Tanaka H, Yamashita S, Hamatani T, Ikemoto Y and Torii S, Chem Lett 1986: 1611
64. Kerdesky FAJ, Ardecky RJ, Lakshmikantham MV and Cava MP (1981) J Amer Chem Soc 103: 1992
65. Hayashi M, Sugiyama M, Toba T and Oguni N, J Chem Soc Chem Commun 1990: 767
66. Vedesj E and Ahmad S (1988) Tetrahedron Lett 29: 2291
67. Czuk R, Glänzer BI, Fürstner A, Weidmann H and Formacek V (1986) Carbohydr Res 157: 235
68a. Kitazume T and Ishikawa N, Chem Lett 1981: 1679
68b. O'Reilly N, Maruta M and Ishikawa N, Chem Lett 1984: 517
68c. Kitazume T and Ishikawa N (1985) J Amer Chem Soc 107: 5186
69a. Pétrier C and Luche J-L (1985) J Org Chem 50: 912
69b. Pétrier C, Einhorn J and Luche J-L (1985) Tetrahedron Lett 26: 1449
70a. Einhorn C, Allavena C and Luche J-L, J Chem Soc Chem Commun 1988: 333
70b. Luche J-L and Allavena C (1988) Tetrahedron Lett 29: 5369
71. Mason TJ, Lorimer JP (1988) Sonochemistry: Theory, Applications and Uses of Ultrasound in Chemistry, Ellis Horwood Chichester

Springer-Verlag and the Environment

We at Springer-Verlag firmly believe that an international science publisher has a special obligation to the environment, and our corporate policies consistently reflect this conviction.

We also expect our business partners – paper mills, printers, packaging manufacturers, etc. – to commit themselves to using environmentally friendly materials and production processes.

The paper in this book is made from low- or no-chlorine pulp and is acid free, in conformance with international standards for paper permanency.

Printing: Saladruck, Berlin
Binding: Buchbinderei Lüderitz & Bauer, Berlin